FAMILY NUTRITION GUIDE

by **Ann Burgess**
FAO Consultant

with **Peter Glasauer**
FAO Food and Nutrition Division

FOOD AND AGRICULTURE
ORGANIZATION OF THE UNITED NATIONS

United Book Prints

United Book Prints

Reprinted in India by:

UNITED BOOK PRINTS
4806/24, Ansari Road, Daryaganj,
New Delhi 110 002, INDIA
E-mail: info@scientificpub.com
Website: www.scientificpubonline.com

Print 2024

ISBN: 978-93-88399-68-5

Printed in India

Preface

Eating well is vital for a healthy and active life. Most people know that we need to eat in order to have the strength to work. However, not everybody has a clear idea about precisely what it means to eat well and how this can be achieved with limited resources. The problem of eating well with limited resources is a particularly important one for many people in developing countries.

Poverty is a major cause of the nutritional problems found in developing countries. But malnutrition also exists where people are not poor and where they can get enough to eat. This is clearly evident in the fact that there are two – quite opposite – main types of malnutrition. The first type is the result of insufficient intake of good-quality and safe foods. The second type is caused by an excessive or unbalanced intake of food or certain types of food. Both can be prevented by an adequate or healthy, balanced diet.

To be well nourished, families need sufficient resources to produce and/or purchase enough food. They also need to understand which combinations of foods make a healthy diet and they need the skills and motivation to make good decisions on family care and feeding practices.

Whether food supplies are scarce or abundant, it is essential that people know how best to use their resources to obtain a variety of safe and good-quality foods. Nutrition education plays a vital role in promoting good nutrition. It is especially important in developing countries where traditional knowledge alone often is no longer enough to deal with the new challenges of rapid and thorough economic and social changes.

Many governments and non-governmental institutions make great efforts to improve people's nutrition, and nutrition education is often one way to do so. To be most effective, nutrition education must apply the latest findings of the nutrition sciences. Also, it must be carried out in a way that truly succeeds in motivating people to adopt healthy diets and lifestyles. Educational programmes need to take into account the advances made in our understanding of nutrition and behavioural change, and the curricula of programmes need to be updated accordingly.

The *Family Nutrition Guide* is a book that can help in this educational process. It provides an up-to-date summary of the relevant nutrition information and gives many suggestions on how to share this information when working with groups of people. The overall purpose of the *Family Nutrition Guide* is to help health professionals in developing countries to provide more effective nutrition education by giving families the information they need to prepare nutritious and safe meals and feed each member of the family well, and by motivating people to adopt healthy eating habits.

The guide is designed primarily for professionals who want to improve the feeding and nutrition of families. It may also be useful to individuals or members of a community group who want to know more about nutritious family feeding.

While the illustrations and food examples in this guide mainly reflect the situation in countries of Eastern and Southern Africa, the basic information in this book is relevant for all regions.

We hope that you, the reader, will find this book useful as a technical guide and that it will help you to design new, or improve existing, nutrition education curricula and material. We also hope that it motivates you to become even more involved in nutrition education. Your opinions are important to us. So we invite you, the user of this guide, to send us your comments on its contents, to share your experiences in its use, and to make suggestions for improving future versions.

FAO is ready to collaborate with governments and institutions that want to improve their nutrition education activities. For example, FAO could help where it is necessary to adapt this *Family Nutrition Guide* to the needs of specific regions and/or communities.

Kraisid Tontisirin
Director
Food and Nutrition Division
FAO

For comments and more information, please contact:
The Director
Food and Nutrition Division
FAO
Viale delle Terme di Caracalla
00100 Rome, Italy
E-mail: nutrition@fao.org
Facsimile: +39 0657053152

Acknowledgements

Many people have contributed to the development of this publication and the authors would like to express their deep gratitude to all of them.

An earlier version of the *Family Nutrition Guide* was prepared under the FAO project TCP/SUD/6714 and benefited greatly from Charity Dirorimwe's broad experience in community nutrition.

During the planning phase of this version of the guide, we obtained very useful practical advice from Anna Mswata, Regional Nutrition Coordinator, Arusha, Tanzania, and Grace Maina, Nutrition Consultant, Nairobi, Kenya. A number of people contributed up-to-date technical information and useful comments, which ensures that the guide takes fully into account both the latest scientific thinking and the experience of those who work in the field of nutrition. For these contributions, we are particularly indebted to: Marlou Bijlsma, University of Zimbabwe, Harare; Bruce Cogill and Ellen Piwoz, Academy for Educational Development, Washington, DC, USA; Andrew Trevett, Cranfield University, UK; Lida Lhotska, IBFAN-GIFA (Geneva Infant Feeding Association), Geneva, Switzerland; Madeleine Green and Andrew Tomkins, Institute of Child Health, London, UK; as well as our colleagues at WHO, Geneva, Switzerland: Peggy Henderson, Constanza Vallenas and Martin Weber, Department of Adolescent and Child Health, and Bruno de Benoist, Randa Saadeh and Catherine Melin, Department of Nutrition for Health and Development.

Within FAO, Ellen Mühlhoff and William D. Clay, Nutrition Programmes Service, provided useful comments and suggestions at various stages of the guide's development. Terri Ballard, Robert C. Weisell and Guy Nantel, Nutrition Planning, Assessment and Evaluation Service, provided technical information on nutrient requirements.

Special thanks go to Sara Kionga-Kamau, Nairobi, Kenya, who prepared both the cover and the illustrations for the key messages in the guide, thereby enriching it considerably.

Several people helped to prepare the guide for printing. Acknowledgements are due to Linda Mitchell for copyediting and coordinating the overall publication process and to Cecilia Valli for document design and layout.

Contents

List of Boxes, Figures and Tables

BOXES

FIGURES

TABLES

Introduction

A. ABOUT THE GUIDE

This guide is for everyone who wants to improve the feeding and nutrition of families in developing countries. It is for *you* if you are a health worker, nutritionist, agricultural extension worker or any other kind of development worker. It is for *you* if you are a member of a community group or a mother or other caregiver who wants to know more about family feeding. It might also be useful to anyone training health staff and community workers.

If you do not have a basic knowledge of nutrition and feel uncomfortable dealing with some technical parts of the guide, we suggest that you team up with local professionals so they can give you help when you need it.

The purpose of the guide is to:

- provide the information needed to prepare good, nutritious and safe meals and feed each member of the family well;
- motivate people to adopt healthy eating habits.

The guide is divided into 11 topics that cover basic nutrition, family food security, meal planning, food hygiene and the special feeding needs of children, women and men, and of old, sick and malnourished people. Each Topic is set out in the same way and has two parts: *Nutrition notes* and *Sharing this information.*

The *Nutrition notes* summarizes up-to-date knowledge on each topic. These can be used to prepare:

- face-to-face education sessions with families and other community-level groups (including teachers, care workers, traditional health workers, etc.);
- nutrition education print materials (such as booklets, brochures, flyers, posters) or material for other media (such as radio talks);
- training materials for different levels of staff in different sectors who deal with family nutrition.

You may also find them useful to update your own and perhaps your colleagues' nutrition knowledge.

Sharing the information is for people working directly with families and community groups. It describes the steps needed to *prepare* an education session. These steps are: finding out the community's present nutritional situation and knowledge; deciding what information to share and with whom to share it; and choosing communication methods. This part also gives some *Examples of questions to start a discussion* which may help to encourage participation and make the session more fun.

Key messages appear throughout the *Nutrition notes* and summarize important points being made in the text. You may use them as 'talking points' or guidelines for structuring a nutrition education session.

The book contains a glossary and three appendixes covering sources of nutrients in foods, energy and nutrient needs, and additional sources of information.

Before using the guide, it is important to adapt the nutrition information to the local area where it will be used. We suggest how to do this in Section B.

B. USING AND ADAPTING THE GUIDE

Diets and eating habits vary from place to place. Families in different areas eat different foods and cook in a variety of ways. They live in different regions where the type and amount of food available can differ considerably. They have different beliefs about foods and how to feed their children. The amount of money, time and other resources they have varies and this affects what they eat. Families differ in what they know about nutrition and they obtain information about nutrition in different ways. Since this guide is written in a 'general' manner and does not reflect a specific country, part of its technical information will always need to be *adapted* so that it is suitable for the areas where it will be used. This will also enrich the guide with local knowledge and experiences. Such adaptation can be seen as regular preparation for the use of the guide and specific guidelines for this process are given below.

Some countries or regions may decide to produce a local (national) version of the guide to make it more focused on their specific situation, regarding food and nutrition problems, type of local foods and eating habits, etc. This will help their national health workers and other users to make good and easy use of the guide. Guidelines for this more thorough process of adaptation, which will result in a new, local version of the guide (or similar materials) are also given below.

Guidelines for using the guide

1. Read the guide. Check the *Nutrition notes* in each topic. Do you understand and agree with the information given? Recent nutrition research means that some of the data may be different from those you have learned before. Consult your supervisor if necessary. You may want to include some information, or change the names of foods, etc., to make the guide suitable for the communities with which you work.

2. Decide, with colleagues, which of the topics are relevant to the local families. This depends on the nutrition problems in your area and whom they affect. You may want to find answers to the following questions. Are many babies born

INTRODUCTION

with a low birthweight? Are many babies not exclusively breastfed? Are poor feeding practices of children or women a problem? Are old or sick people fed poorly? Is anaemia a problem and who is most affected? Are many people overweight? What are the nutritional and other causes of these disorders? Do people living with HIV/AIDS know which foods help them to stay well? What nutrition information do groups and families request themselves?

3. Select the topic(s) you want to share and decide how to do this. Unless you are sure that people understand the basic facts of healthy nutrition, you should try to include Topics 1, 2, 3 and 4 in any nutrition education (or training) course. Box 1 suggests a good order in which to use the Topics. Suggestions for how to select the Topics in different situations are given in the following examples.

   - You may be working with a group or family on several occasions. For example, you may be making several visits to a youth or religious group or you may be working with mothers and caregivers who regularly come to a young child or antenatal clinic, or to community growth monitoring sessions. In this case, you may have sufficient time to follow the order suggested in Box 1. If you have time to cover only a few topics, start with Topic 1 to 'set the scene' and then choose only those topics that are relevant to and/or requested by the group or family.

BOX 1 • SUGGESTED ORDER FOR SHARING THE TOPICS

TOPIC 1 WHY WE NEED TO EAT WELL → **TOPIC 2** GETTING ENOUGH FOOD → **TOPIC 3** MAKING GOOD FAMILY MEALS → **TOPIC 4** KEEPING FOOD SAFE AND CLEAN

↓

Then choose one or more or all of the following:

- **TOPIC 5** FOOD AND CARE FOR WOMEN
- **TOPIC 6** FEEDING BABIES AGED 0–6 MONTHS
- **TOPIC 7** FEEDINGYOUNG CHILDREN AGED OVER SIX MONTHS
- **TOPIC 8** FEEDING SCHOOL-AGE CHILDREN AND YOUTHS
- **TOPIC 9** FEEDING MEN AND OLD PEOPLE
- **TOPIC 10** FEEDING SICK PEOPLE
- **TOPIC 11** PREVENTING AND MANAGING MALNUTRITION

- You may be invited by a group of women or farmers to talk about one specific topic, such as feeding children. In this case you may need to use parts of Topic 1 (Why we need to eat well), Topic 3 (Making good family meals) and/or Topic 4 (Keeping food safe and clean) so that participants understand the basics of nutrition, meal planning and hygiene. Then use Topic 6, 7 or 8, each of which discusses in full about feeding children of different ages.
- You may be visiting the home of a malnourished child. This gives you a chance to assess the particular educational (as well as other) needs of the family. You may need to use parts of Topic 1 (if you think the mother or other caregiver needs a reminder of 'basic nutrition') and perhaps Topic 4 (if poor hygiene is a problem) and then Topic 11 and – depending on the child's age – Topic 7 or 8.

4. Select information from the *Nutrition notes* that is relevant and practical for the particular group or family. *Do not try to cover too much at one time.* It is better to share a *small amount of relevant* information than to cover all the material in the Topic. Adapt the information and advice to the situation and resources of the group or family (e.g. suggest local foods that a family can produce or buy; demonstrate recipes using local stoves and fuel; use local words for disorders such as anaemia).

5. Select the method for sharing the information. This depends on *whom* you are trying to reach. Some examples are the following.
   - Group discussions. These are useful at clinics and meetings of community groups, such as women's, youth, farmers' and religious groups. Box 2 explains how to use questions to encourage participation and make the discussion livelier.
   - Counselling of individuals or families. This can be done in private, at clinics, in maternity units or in homes. Counselling is a dialogue between you and another person (e.g. a mother, a father) which helps the person make informed decisions about her/his future behaviour.
   - Demonstrations (see Figure 1, page 6). Demonstrations are useful to show how to prepare a meal or snack, how to keep food hygienically and how to feed a young child or sick person.
   - Personal example. In most communities there are people who, in spite of limited resources, are feeding their families successfully. A good way to persuade other families to improve food and feeding practices is to ask these successful people to share information about what they are doing. For example, mothers who exclusively breastfeed can talk with pregnant women; families with healthy young children might explain how they sit with their children at meals and encourage them to eat; schools with successful gardens can share knowledge of gardening methods with other schools; women's groups can share recipes for preserving foods.
   - Songs, poems and drama. Use these to introduce a topic or reinforce messages.

6 Find the blocks that may prevent families from improving their feeding practices. These may be: lack of resources (such as money, women's time); existing beliefs, traditions and food taboos; pressures from other family members, particularly men; unavailability of foods or agricultural inputs; inappropriate or culturally insensitive advice.

If an individual or family is not feeding as recommended, find out why. There may be many reasons and you may have to probe sensitively to discover them (e.g. the family may be embarrassed by lack of money; a mother is not

BOX 2 • DISCUSSION GROUPS – HOW TO ENCOURAGE PARTICIPATION

One way to make a discussion more interesting and effective is to *ask questions* instead of telling people what to do. Asking questions allows you to find out what people already know and believe. It makes people take an active part and discuss together traditional beliefs and new concepts. It is more likely to lead eventually to small or large changes in behaviour.

Example of the start of a discussion on feeding during pregnancy (see Topic 5)

Situation: A meeting of a women's group in a rural area of Africa. The group identified the problem that some women do not eat extra food during pregnancy. You, a nurse, have been invited to the meeting to lead the discussion. Here are some questions that you can use to stimulate participation. You may need to encourage some people, especially shy or young women, to take part. You may have to prevent one or two people from dominating the meeting. Remember to be very polite. Never tell anyone they are 'wrong' but emphasize 'correct' responses and beliefs. Aim to get a group to agree to make at least a small change to improve any poor feeding practices.

Ask: "Thank you for inviting me to this meeting so we can discuss together how to eat well during pregnancy. First, let me ask, should we eat more or less food when we are pregnant?" Wait for a few responses. Then say, "I agree with those of you who said 'more', it is important to eat more during pregnancy".

Ask: "Why do pregnant women need extra food?" Agree with correct responses: for example, "We need extra food to build the placenta and the baby's body". Add other reasons if they are not mentioned. Gently correct 'wrong' answers if necessary. Explain that women who do not eat enough are likely to have small, weak babies.

Ask: "How can pregnant women increase their food intake during pregnancy?" Responses could include 'eating more often' or 'eating more at each meal'. Describe or show an example of the amount of extra local foods needed.

Ask: "Are there any foods that are especially good for pregnant women?" Allow several responses and catch 'correct' ones (e.g. iron-rich foods, such as beef or liver).

Continue asking questions and 'catching' and explaining correct answers. Give people plenty of opportunity to ask you questions. At the end say, "Are there any other questions? Thank you. I have learned a lot from you today".

breastfeeding because she fears she is HIV+). Then discuss together what a family is able and willing to do. It may be best to first encourage a small, easy change in behaviour. A family may agree to make a small change but not a big one. For example, a mother may agree to spend more time feeding a young child but would not be able to give the child an extra meal a day. Discussions with other family members may help a mother make a change. Perhaps the family can do some of the mother's work so she has more time to feed a sick child.

7. Evaluate your work. The purpose of the guide is to help families improve feeding practices. You may want to interview families or groups to ask what information they found useful, what they learned that was new and what they have put into practice. When you see an individual family again, find out if they have made any of the suggested behavioural changes. If not, try to find out why. This will help you to modify the information you share and the way you share it. You may need to reinforce advice given by presenting it in different ways. Make sure that you and your colleagues are giving the same nutritional messages.

Figure 1. Demonstrating how to prepare a good meal

Guidelines for adapting the guide

Adapting the guide:

- makes the information relevant to local families and local nutrition problems;
- provides an opportunity for nutritionists and others from different sectors and organizations to discuss the material and to update national or local nutrition guidelines. This process helps to create a sense of ownership of the guide.

Adaptation of the guide can be done at national or provincial level. People who might be able to help to prepare and produce a local version of the guide include:

- staff from a regional nutrition institute or from departments of home economics or of food science and technology;
- an experienced nutritionist, dietician and/or medical doctor who is familiar with the area and its problems;
- staff from an appropriate United Nations agency, such as the Food and Agriculture Organization (FAO), the World Health Organization (WHO) and the United Nations Children's Fund (UNICEF).

1. Before you and your colleagues adapt the guide you should discuss, find out and decide the following.
   - Which 'nutrition educators' will use the guide. Will they be trained nurses and other health staff? Agricultural extension or other development workers? Community health workers? Leaders of community groups? Literate parents responsible for feeding the family? You need to know the level of knowledge and education of these educators so you make your guide relevant and easy for them to use. You may need to translate the adapted guide (or only the main messages) into the local language. You also need to know in which situations the guide will be used and what other learning/teaching materials the educators have.
   - The characteristics of families with whom you or other educators will work. You need to know:
     - what foods are available and eaten at different seasons; what the local feeding practices, knowledge and beliefs are; what local recipes are used; how food is shared among family members; how babies, young children, older children, women, men and old people are fed; what resources are available (e.g. land, money, water, time); how food is produced, stored, preserved and cooked; what the levels of hygiene are; which foods are eaten outside the home and by whom; who makes decisions related to family feeding;
     - what types of nutrition problems and malnutrition are found in these families; which family members are at risk and what the probable causes are; what peoples' perceptions and beliefs are, as well as their explanations about nutritional disorders and their causes (see Section C).

2. Decide which parts in the original guide to alter or delete, and what information to add. You will probably want to make changes to or add information on:
   - the feeding problems and types of malnutrition found locally;
   - foods suggested for healthy, balanced diets, adding important local foods that are not mentioned in the guide;
   - words used for foods, recipes, measures of weight and volume, etc., using local words;
   - methods used to increase food production;
   - varieties of legumes, vegetables and fruits to grow and animals and fish to raise;
   - methods used to store and preserve food, mentioning practices that can contaminate and/or waste food and making suggestions for improved practices;
   - ways used to preserve nutrients during cooking, emphasizing local methods that should be encouraged and pointing out those that decrease nutrient values;
   - advice on budgeting and 'good buys';
   - local recipes for feeding young children and sick people;
   - food composition tables, trying to use local ones;
   - sources of more information;
   - illustrations, making sure that any changes or additions are culturally as well as technically correct.
3. Prepare the revised guide. Do not make it too long and only include information that educators need to help families have good, balanced diets.
4. Field-test the guide with some educators and target families. Is the information practical as well as technically and culturally correct? Is the advice feasible for the different types of families you want to reach? Do the educators understand the text, concepts and illustrations? Is your guide easy to read and use?
5. If changes are still necessary, revise the guide, test it again and then prepare a final version.

C. WHAT HAPPENS IF FAMILIES DO NOT EAT WELL

The consequences of not eating well

(also see Topic 11)

People who have poor diets and do not eat the right amounts of energy-rich food and nutrients are often sick and become malnourished. The type of malnutrition that occurs depends on which nutrients and how much of the required food energy are lacking (or are in excess) and for how long, and the age of the person.

1. Children and adults may eat too little food and become undernourished because they do not have enough food or they have a poor appetite. These people lack energy and many nutrients, which means:
   - they have less energy so they cannot work, study or play as normal;
   - their immune systems are weak so they become ill easily and/or are seriously ill;
   - children stop growing and may lose weight. If very little food is eaten (often because of infection), a child may develop severe malnutrition (i.e. kwashiorkor or marasmus);
   - adults lose weight. If a pregnant woman is undernourished, her unborn baby grows poorly.

2. People may eat unbalanced diets that provide too little of a particular nutrient. For example:
   - if there is a lack of iron, the mental and physical development of children may be delayed. People of all ages are less active, have less immunity to infections and may become anaemic. Anaemic women have an increased risk of dying during and after pregnancy;
   - if there is a lack of iodine, people become more apathetic and find it hard to work or study. Sometimes a goitre develops. A woman who lacks iodine in early pregnancy is at high risk of having a child who is mentally and physically damaged. For example, the child may have a lower IQ or be deaf;
   - if there is a lack of vitamin A, people are more likely to become sick because the immune system is damaged. In severe vitamin A deficiency, there are eye conditions that range from night blindness to dry eyes (xerophthalmia), to corneal damage and blindness. These eye conditions occur most often in young children and pregnant women.

3. People may eat more food (especially energy-rich foods with plenty of fat and/or oil) than they need. By taking in too much energy they become too fat

(overweight or obese). These people are at increased risk of chronic conditions, such as heart disease, high blood pressure and diabetes (see "Overweight and obesity", page 14).

Malnutrition (due to both lack and excess of food energy and/or nutrients) is one of the biggest health problems in the world, especially in developing countries.

- More than half the deaths of children aged 0–5 years are associated with undernutrition.
- In many countries a third of the young children are stunted and 10 percent are too thin (wasted).
- About a sixth of newborns have low birthweights, which makes them more likely to become ill, grow slowly and die.
- Anaemia caused by lack of iron is the biggest nutrition disorder. In many places half the women are anaemic.
- Vitamin A, iodine and zinc deficiency disorders are widespread in many countries.
- Overweight and obesity and their related disorders are on the increase in most countries (see "Overweight and obesity", page 14).

Causes of malnutrition

There are many reasons why a child or adult becomes undernourished. The causes vary from person to person but we can divide them into immediate, underlying and basic causes.

Immediate causes

These are a poor diet and disease.

- A poor diet may be due to:
  - insufficient breastmilk;
  - meals that are too small;
  - poor variety of food;
  - low concentrations of energy and nutrients in meals (i.e. food is too 'watery');
  - infrequent meals.
- Disease. Sick people may:
  - not eat much;
  - absorb few nutrients;
  - lose nutrients from the body;
  - use up nutrients in the body more quickly (e.g. during fever).

Underlying causes

These include family food shortages, inadequate care and feeding practices, especially of children and women, and poor living conditions and poor health services.

- Family food shortages, which may be due to:
  - lack of money for food;
  - low production of family food;
  - poor food storage and preservation;
  - poor choices and budgeting.
- Inadequate care and feeding practices:
  - the way families feed young children and encourage them to eat;
  - the way families care for women (especially during pregnancy, childbirth and breastfeeding) and for sick and old people;
  - the way food is prepared and the level of hygiene in the home;
  - the ways families prevent and treat illnesses at home and use health facilities.
- Poor living conditions (e.g. insufficient water, inadequate sanitation and over-crowded housing) and poor health services. Shortages of medicines and skilled health staff increase the risk of disease. Inadequate environmental sanitation services increase the risk of food-borne infections.

The role of women in food production, trade and preparation is vital but is often overlooked when causes of malnutrition are analysed and nutrition programmes are planned. In many countries, women produce much or most of the food. The level of care and quality of diet that women can give their families (including themselves) depends largely on their workloads and their social role within the family. For example, when women have heavy workloads (which many do), they may not have time to prepare more than one meal a day (which is especially insufficient for young children). If women have little authority and little control over resources (e.g. land, money), this also affects the type of care they can give different family members. Women's workloads and social roles can be important underlying causes of malnutrition.

Basic causes

For each underlying cause there are 'deeper' causes. These may include:

- widespread poverty and lack of employment opportunities;
- unequal distribution and control of resources at community, district, country and international levels;
- the low status and education of women;
- population pressures;
- environmental damage;

- political unrest and conflict;
- lack of health, education, and other social services;
- discrimination.

Figure 2 on the opposite page demonstrates many of the different factors at various levels of society that can lead to malnutrition. Of course, these factors are more complicated in real life. In fact, malnutrition itself can reduce the ability of a family to care for all its members – and so creates a vicious circle of malnutrition and its underlying causes. For example, this happens when a malnourished child needs more attention from caregivers and hence further weakens the family's capacity to look after the needs (food, health, etc.) of other family members. The series of illustrations in Figure 2, nevertheless, helps us to identify the most important reasons why a person, family or community can be malnourished.

Box 3 below shows how HIV/AIDS is both an immediate and underlying cause of malnutrition (also see Topic 10, page 84).

BOX 3 • HOW HIV/AIDS CAUSES MALNUTRITION

HIV/AIDS can be an *immediate* cause of malnutrition because:

- a person living with HIV/AIDS:
    - may have a poor appetite, sore mouth or nausea and so eats less;
    - absorbs fewer nutrients due to diarrhoea and a damaged gut;
    - uses nutrients faster because the immune system is working harder than usual.
- an HIV+ mother who decides not to breastfeed may be unable to give adequate replacement feeds.

HIV/AIDS can be an *underlying* cause of food insecurity and malnutrition if:

- productive adults become sick and are less able to farm or earn money for food, or if they die;
- a family sells its assets (e.g. cattle, tools) to get money for food and medicines;
- family members stop farming or paid work in order to care for a sick relative;
- orphaned children have to care for younger brothers and sisters. These children often lack the skills and resources to produce enough food or provide good meals. They may have to stop going to school and so reduce their chances of a good job in the future;
- old people who have to care for young orphans do not have the energy or money to feed them well.

Children often become malnourished if one or more parents is sick or dead. They may lack food and care, or they may eat less because of grief and depression.

Adapted from the UNICEF Framework of Underlying Causes of Malnutrition and Mortality

Figure 2. Immediate, underlying and basic causes of malnutrition

Overweight and obesity

Overweight and obesity are principally caused by regularly:

- eating too much food, particulary energy-rich food (often containing large amounts of fat and sugar);
- having a lifestyle (work, sports, travel) that does not involve enough physical activity.

Obesity is now a worldwide epidemic. More than 1 billion adults are overweight and at least 300 million adults are obese (see Glossary for definitions of overweight and obesity). Overweight and obesity affect almost all ages and socio-economic groups, and the increasing numbers of obese children is a major cause for concern. In many developed *and* developing countries the obese proportion of the population has increased threefold since 1980. And this increase is often faster in developing countries than in developed ones. In most developing countries obesity *and* undernutrition are now major problems.

The increase in obesity is one of the main reasons for the increase in diet-related chronic disease and disability.

- The non-fatal, but debilitating health problems linked to obesity include respiratory difficulties, chronic musculoskeletal problems, skin problems and infertility.
- The life-threatening problems are cardiovascular diseases, including hypertension and stroke, type 2 diabetes, certain types of cancers, and gallbladder disease.

The health consequences of these conditions range from premature death to disabilities that reduce the quality of life.

TOPIC 1 WHY WE NEED TO EAT WELL

NUTRITION NOTES

Good meals

Eating good food, especially with family and friends, is one of the pleasures of life. We all know that people who eat healthy, balanced diets are likely to have:

- plenty of energy to work and enjoy themselves;
- fewer infections and other illnesses.

Children who eat well usually grow well. Women who eat well are likely to produce healthy babies. That is why it is important to know which combinations of foods make good meals and what the different food needs of different members of the family are.

Foods and nutrients

Foods provide nutrients so we can grow and be active and healthy

A food is something that provides nutrients. Nutrients are substances that provide:

- energy for activity, growth, and all functions of the body such as breathing, digesting food, and keeping warm;
- materials for the growth and repair of the body, and for keeping the immune system healthy.

There are many different nutrients. We divide them into:

Macro (big) nutrients that we need in large amounts. These are:

- carbohydrates (starches, sugars and dietary fibre);
- fats – there are several kinds (see Box 4);
- proteins – there are hundreds of different proteins.

Micro (small) nutrients that we need in small amounts. There are many of these but the ones most likely to be lacking in the diet are:

- minerals – iron (see Box 6, page 19), iodine and zinc;
- vitamins – vitamin A, B-group vitamins (including folate) and vitamin C.

BOX 4 • FATS, FATTY ACIDS AND CHOLESTEROL

Fats and oils provide a concentrated source of energy and the essential fatty acids needed for growth and health. They aid the absorption of some vitamins such as vitamin A and improve the taste of meals. Some fatty/oily foods contain important vitamins.

Fats and oils contain different 'fat-nutrients'. These include unsaturated fatty acids, saturated fatty acids, trans fatty acids and cholesterol.

Unsaturated fatty acids

Two of the unsaturated fatty acids are called 'essential fatty acids' because the body cannot make them. They are needed for building cells, especially the cells of the brain and nervous system. Unsaturated fatty acids contain a group called 'omega-3 fatty acids', which help to protect the body from heart disease.

- Examples of foods containing mainly unsaturated fatty acids are most vegetable oils, groundnuts, soybeans, sunflower seeds, sesame seeds and other oilseeds, oily fishes and avocados. Foods rich in omega-3 fatty acids are oily sea fish and some seeds and pulses such as linseed and soybeans.

Saturated fatty acids

- Examples of foods containing mainly saturated fatty acids are butter, ghee, lard/cooking fat, whole milk, cheese, fats from meats and meat products (e.g. sausages) and poultry, red palm oil and coconuts.

Trans fatty acids

When vegetable oils are processed to make them harder (e.g. for use in margarine and other solid fats), some of the unsaturated fatty acids are changed into trans fatty acids. These behave like saturated fatty acids. We should eat as little of the foods containing trans fatty acids as possible.

- Examples of foods containing trans fatty acids are margarine and lard (shortening), fried foods, such as chips (French fries) and others, commercially fried foods, such as doughnuts, as well as baked goods, biscuits, cakes and ice creams.

Cholesterol

Cholesterol is found only in animal foods but the body can make it from other fat-nutrients. We need some cholesterol for our bodies to grow and function properly.

There are two kinds of cholesterol in the blood.

- High levels of 'good' cholesterol (high-density lipoprotein) seem to *reduce* the risk of heart disease. Eating foods containing mainly unsaturated fatty acids tends to *increase* the level of good cholesterol.
- High levels of 'bad' cholesterol (low-density lipoprotein) seem to increase the risk of heart disease. Eating foods containing mainly saturated fats tends to increase the level of bad cholesterol.

Fat needs

Fat needs are expressed as 'percent of total energy needs' (see Appendix 2). The percent of total energy that should come from fat in a healthy balanced diet is:

- 30–40 percent for children on complementary feeding and up to the age of two years;
- 15–30 percent for older children and most adults; for active adults up to 35 percent is acceptable;
- At least 20 percent up to 30 percent for women of reproductive age (15–45 years).

This means the diet of a woman of reproductive age who needs approximately 2 400 kcal/day should contain about 480–720 kcal from fat or oil. This is equivalent to 53–80 g of pure oil (or about 11–16 level teaspoons). Part of the fat in a diet is not added in the kitchen at home but is 'hidden' in foods such as meat, milk, groundnuts and oilseeds as well as fried foods.

Fat and health

It is recommended that less than one-third of the fat in the diet is in the form of saturated fatty acids. Red palm oil and coconuts/coconut oil are foods rich in saturated fatty acids but, unlike other such foods, they do not seem to increase the risk of coronary heart disease. Moderate intake of coconut, for instance, seems to be acceptable, providing other foods high in saturated fats are eaten as little as possible. This is particularly true where the overall lifestyle lessens the risk of heart disease. Such a lifestyle could, for example, be one with a high physical activity level, high intake of fish, vegetables and root crops, low intake of salt and little or no use of tobacco or alcohol. Red palm oil is also a good source of other important nutrients, such as vitamin A and vitamin E.

Ideally trans fatty acids should provide less than 1 percent of the total energy intake (or not more than 2 g for most adults).

For many families this means they should, when possible, eat *more* of the foods rich in unsaturated fatty acids (e.g. foods from plants and oily sea fish), *less* of the foods high in saturated fatty acids, and *much less* of the foods high in trans fatty acids.

Foods rich in unsaturated fatty acids are better for the health of the heart than foods high in saturated or trans fatty acids

Our bodies use different nutrients in different ways as shown in Box 5.

BOX 5 • IMPORTANT USES OF SOME NUTRIENTS

Nutrient	Main use in the body
Macronutrients	
Carbohydrates – starches and sugars	To provide energy needed to keep the body breathing and alive, for movement and warmth, and for growth and repair of tissues. Some starch and sugar is changed to body fat.
Carbohydrates – dietary fibre	Fibre makes faeces soft and bulky and absorbs harmful chemicals, and so helps to keep the gut healthy. It slows digestion and absorption of nutrients in meals, and helps to prevent obesity.
Fats	To provide a concentrated source of energy and the fatty acids needed for growth and health. Fat aids the absorption of some vitamins such as vitamin A.
Proteins	To build cells, body fluids, antibodies and other parts of the immune system. Sometimes proteins are used for energy.
Water	To make fluids such as tears, sweat and urine, and to allow chemical processes to happen in the body.
Micronutrients	
Iron	To make haemoglobin, the protein in red blood cells that carries oxygen to the tissues. To allow the muscles and brain to work properly.
Iodine	To make thyroid hormones that help to control the way the body works. Iodine is essential for the development of the brain and nervous system in the foetus.
Zinc	For growth and normal development, for reproduction and to keep the immune system working properly.
Vitamin A	To prevent infection and to keep the immune system working properly. To keep the skin, eyes and lining of the gut and lungs healthy. To see in dim light.
B-group vitamins	To help the body use macronutrients for energy and other purposes. To help the nervous system to work properly.
Folate	To make healthy red blood cells and to prevent abnormalities in the foetus.
Vitamin C	To aid the absorption of some forms of iron (see Box 6). To destroy harmful molecules (free radicals) in the body. To help wound healing.

BOX 6 • IRON ABSORPTION

Iron from meat, liver and other offal, poultry, fish and breastmilk is well absorbed in the gut. Iron from other milks, eggs and all plant foods is poorly absorbed, but other foods in the same meal affect the absorption of this type of iron.

- Meat, fish and vitamin C-rich foods (fresh fruits and vegetables) *increase* the absorption of this type of iron so *more* is absorbed.
- Some foods, such as tea, coffee and wholegrain cereals, contain 'antinutrients' (e.g. phytate) that *decrease* the absorption of this type of iron.

The best way to make sure that we get enough of each nutrient and enough energy is to eat a *mixture* of foods. Topic 3 explains how to combine foods to make good meals. Appendix 1 lists sources of each nutrient (see Tables 1 and 3) and the nutrient content of different foods (see Table 2).

Different types of foods

Different foods contain different mixtures of nutrients

Staple foods are usually cheap and supply plenty of starch (for energy), some protein, some micronutrients (especially some of the B-group vitamins) and dietary fibre.

Circle the staple foods used locally and add others to the list.

CEREALS	STARCHY ROOTS AND FRUITS
MAIZE/CORN MEAL	POTATO
MILLET	SWEET POTATO
SORGHUM	YAM
WHEAT FLOUR	FRESH CASSAVA
RICE	CASSAVA FLOUR
	PLANTAIN

TOPIC 1

Legumes and oilseeds. Legumes are good sources of protein, some micronutrients and dietary fibre. High fat legumes and oilseeds provide fat.

Circle the legumes and oilseeds used locally and add others to the list.

LOW FAT LEGUMES	HIGH FAT LEGUMES AND OILSEEDS
COWPEA	GROUNDNUT
PIGEON PEA	SOYBEAN
KIDNEY BEAN	PUMPKIN SEED
LENTIL	SUNFLOWER SEED
CHICKPEA	SESAME (SIM SIM)
	MELON SEED

Milk. Breastmilk can supply all the nutrients needed for the first six months of life and a useful proportion of the nutrient needs up to at least 2 years of age. Animal milks, and milk foods, such as curds, yoghurts and cheese, are excellent sources of protein, fat and many micronutrients, such as calcium (but not iron).

Eggs are a good source of protein and fat and several micronutrients.

Meat, poultry, fish and offal from these foods are excellent sources of protein and often of fat. They supply important amounts of iron (especially red meat and red offal) and zinc, and many other micronutrients including some B-group vitamins. Liver of all types is a very rich source of iron and vitamin A.

The best sources of iron are meat, offal, poultry and fish

Circle the animal foods used locally and add others to the list.

MILK AND EGGS	MEAT, OFFAL, POULTRY AND FISH
FRESH COW MILK	BEEF
FRESH GOAT MILK	GOAT
CURDS	SHEEP
YOGHURT	PORK
CHEESE	LIVER
	OTHER OFFAL
EGGS	CHICKEN
	FRESH FISH
	DRIED FISH
	CANNED FISH

Fats and oils are concentrated sources of energy. For example, one spoon of cooking oil contains twice as much energy as one spoon of starch or one spoon of sugar. Fats contain fatty acids some of which are needed for growth. In addition to 'pure' fats (e.g. butter) and 'pure' oils (e.g. maize oil), other rich sources of fats and oils are oilseeds, cheese, fatty meat and fish, avocados and fried foods. Red palm oil is a rich source of vitamin A.

Sugar gives only energy and no other nutrients. It is useful for making foods taste nice and for improving appetite, for instance during illness. However, eating sugary foods too often can be harmful to health for several reasons. Sweet, sticky foods, such as ice lollies, or snacks and pastries prepared with pleanty of sugar, honey or syrup, are bad for the teeth if eaten often. Many sugary foods also contain much fat, which increases the risk of 'overeating' for those who should limit their energy intake. People who often eat sugary foods and consume sweet drinks such as sodas (bottled fizzy drinks) are more likely to become overweight and to develop diabetes. These people also often eat less of other, more nutrient-rich foods. There is much sugar in sweets (candy), lollies, sodas, jam and sweet cakes and biscuits.

Eating too much sweet sticky food is bad for the teeth

Eating too much sugary food often means eating less of other, more nutrient-rich foods

Circle the fats, oils and sugars used locally and add others to the list.

FATS	OILS	SUGARS
BUTTER	COCONUT OIL	TABLE SUGAR
MARGARINE	SUNFLOWER OIL	HONEY
FAT FROM MEAT	GROUNDNUT OIL	JAM
GHEE	RED PALM OIL	SUGAR CANE
	MAIZE OIL	

Vegetables and fruits are important sources of micronutrients and dietary fibre but the amounts vary according to the type of vegetable or fruit. Orange vegetables, such as orange sweet potato and carrots, and orange fruits, such as mango and pawpaw but not citrus fruits (e.g. oranges and lemons), are excellent sources of vitamin A. Most fruit and fresh (not-overcooked) vegetables provide vitamin C. Dark green vegetables supply folate and some vitamin A. Many vegetables (e.g. tomatoes, onions) provide additional important micronutrients that may protect against some chronic conditions such as heart disease. The best way to make sure we get enough of each micronutrient and enough fibre is to eat a *variety* of vegetables and fruits every day.

Circle the vegetables and fruits used locally and add others to the list.

VEGETABLES	FRUITS
TOMATO	MANGO
PUMPKIN	PAWPAW
CARROT	PASSION FRUIT
SPINACH	ORANGE
OTHER LEAVES	PINEAPPLE
ORANGE SWEET POTATO	BANANA
SWEET PEPPER	DATES
EGGPLANT	
OKRA	

Flavouring foods. Everyone uses salt in cooking and there is salt in many processed foods. Too much salt is harmful and can lead to high blood pressure. Iodized salt is an important source of iodine. Herbs, spices, garlic and onions are examples of other flavouring foods that help to make meals tasty.

Water. We need about eight cups of water each day and more if we are sweating or have a fever or diarrhoea. In addition to drinking water, we get water from tea, coffee, juices and soups, and from fruits and vegetables.

Food needs of the family

The amounts of different nutrients a person needs varies with age, sex, activity and whether menstruating, pregnant or breastfeeding. Needs also vary during sickness and recovery. The nutrient needs of different family members are listed in Appendix 2, Table 4, and are discussed in Topics 3 and 5–10.

SHARING THIS INFORMATION

Before sharing this information with families, you may need to:

1. **Find out.** What different types of local foods are eaten. What people already know about foods and nutrients.
2. **Prioritize.** Decide which information is *most important* to share with groups or individual families.
3. **Decide whom to reach.** For example: parents and other caregivers, teachers, older school children, youths and leaders of community groups.
4. **Choose communication methods.** For example: illustrated talks, discussions, and demonstrations of foods.

Examples of questions to start a discussion

(choose only one or two questions that deal with the information families need most)

Why do we need to eat well?

Can you list some important minerals and vitamins?

Which important nutrients are found in: cereals, legumes, milks, meats?

Why is too much sugar bad for us?

Why is the fat in plant foods usually more healthy than the fat in margarine or street foods?

Why do we need iron? Which foods are the best sources of iron?

TOPIC 2 GETTING ENOUGH FOOD

NUTRITION NOTES

Food security

BOX 7 • FAMILY FOOD SECURITY

A family is food secure if it has sufficient safe and nutritious food throughout the year so that all members can meet their nutrient needs with foods they like/prefer for an active and healthy life.

People usually get food by producing or buying it. Sometimes they gather wild foods. In times of food shortages they may receive free or subsidized food. To be food secure, people need *enough* food and a *variety* of foods.

- Signs that a family is short of food include: people say they are hungry; they eat fewer meals or smaller than usual meals each day; children grow slowly and/or there is little food in the home.
- Signs that a family has little variety in their diet are: the same few foods are eaten daily; the family eats few vegetables or fruits or little food from animals; and/or they say they have a dull monotonous diet.

Families may become more food secure if you help them to improve:

- food production and storage;
- food preservation;
- food budgeting;
- incomes.

Improving food production and storage

Family farmers may be able to increase the amount and types of foods they produce by:

- improving farming methods (e.g. mulching, composting, intercropping, fertilizing, including use of green manure);

- joining cooperatives to buy fertilizer or other agricultural resources;
- harvesting water for small-scale irrigation;
- using higher yielding seeds or growing crops that mature early or are drought resistant;
- increasing the variety of foods grown, especially vegetables and fruits.

Some rural families may be able to make fish ponds or raise small animals (e.g. poultry, rabbits). Pastoralists may be able to get more productive breeds of animals or learn how to care better for them. Fishermen may be able to increase catches by using better fishing methods.

Figure 3. Families may be able to produce more food

Even people with small amounts of land may be able to improve kitchen gardens or grow vegetables in containers. Refer families that need help to the relevant extension services or to successful local farmers and fishermen. Also see FAO. 2001. *Improving nutrition through home gardening. A training package for preparing field workers in Africa* (listed in Appendix 3).

Improving stores reduces losses of harvested food crops

Much food is lost in on-farm storage. Improved secure stores and safe use of pesticides increase the amount of food available. Sometimes community stores are a good way for farmers to store their crops and seeds. Ask an agricultural extension worker to give families information on better storage if they need it. Food storage in the kitchen is discussed in Topic 4.

Improving food preservation

Some foods can be preserved so they keep longer (e.g. by drying). If necessary, show families practical methods for preserving foods, such as drying vegetables, fruits or fish. Or ask a home economics colleague to demonstrate food preservation methods.

Flour, porridges and milks keep longer if they are soured or fermented. This also improves the digestion of these foods and increases the absorption of iron from the food.

Improving budgeting for food

Find out which foods give the best 'value for money'

Some families need advice on how to budget for food and how to use their money in a more efficient way. They may need to know which foods give *value for money*. This depends on the prices of available foods and this may vary with season, type of shop, etc. To be able to advise families on which foods may be 'good buys' in your area:

- look in Appendix 1 (see Tables 1, 2 and 3) at the lists of foods that are useful sources of different nutrients;
- then compare the prices of similar foods (e.g. different legumes, different iron-rich foods) in different shops and markets (see Figure 4, page 28).

Remember that different foods have different amounts of waste (skin, bones) and some may be adulterated (e.g. milk diluted with water; spices mixed with ground up bricks or stones).

Buying food in large quantities may save money. Most families do not have the money or storage space to do this, but sometimes a group of families can buy in bulk and share the food (e.g. beans, sugar).

Figure 4. Finding out which foods are good value for money

Food that is of *poor quality* is poor value for money. Box 8 below lists signs of poor-quality food.

BOX 8 • SIGNS OF POOR-QUALITY FOOD

Cereals and other dry foods	Contain insects or dirt; look or smell damp or mouldy; bag is broken; legumes are wrinkled; flour is lumpy.
Roots	Soft, sprouting, bruised or damaged; rotten spots.
Vegetables and fruits	Wilted, too soft, rotten spots, bruised.
Meat, poultry and fish	Bad smell or colour; fish have dull eyes or loose scales. Uninspected meat, liver and other offal may contain dangerous parasites.
Fresh milk	Smells bad; is, or has been, exposed to dirt and flies.
Canned foods	Can is swollen, rusty or damaged; food has leaked out; food looks, smells or tastes bad. Any of these signs means the food may be very poisonous.

Advise people to check 'sell by' (and 'use by') dates on labels and not to buy (or use) foods after these dates.

Source: Adapted from Burgess and others. 1994. *Community nutrition for Eastern Africa.* AMREF, Nairobi.

Some foods are poor value for money because they contain *few nutrients* and are expensive. Examples are sodas (bottled fizzy drinks), ice lollies and sweets, which are mainly sugar and so are bad for the teeth (see Topic 1, page 21). These foods should be kept as treats and not given often to children.

Foods fortified with micronutrients are often *'poor buys' especially if they cost a lot.* Exceptions to this rule are salt and fortified staple foods (cereal flours). Usually they do not cost much more than the non-fortified variety and can therefore bring some nutritional advantage at an acceptable cost.

Buy iodized salt if available

In general, it is best to obtain nutrients by eating a healthy diet. Buying a food fortified with a micronutrient is only justified if there is a serious lack of foods containing that particular micronutrient. An example is iodized salt. Unless people can regularly eat fish and other foods from the sea (which are rich in iodine), they are likely to develop iodine deficiency. This is because soils in many parts of the world, and the plant and animal foods raised on them, are low in iodine (Section C explains what happens if people lack iodine). Iodine deficiency disorders are serious and widespread and so, in most places, iodized salt is more than a good buy – it is a 'must buy'.

Nutrient supplements and 'tonics' are usually poor buys. They are often expensive and we should get the nutrients we need by eating a variety of different foods.

Gathering wild foods

Wild foods increase the variety of foods in the diet – and make meals more tasty. The list below gives some examples of useful, nutritious wild foods.

Circle the wild foods used locally; add others, local names and methods of preparation to the list.

WILD FOODS	LOCAL NAMES	WAYS TO PREPARE
BAOBAB LEAVES		
WILD FIGS		
OYSTER NUTS		
BERRIES		
MICE/RATS		
SMALL BIRDS		
CATFISH		
SHELLFISH		
LOCUSTS		
FLYING ANTS		

SHARING THIS INFORMATION

Before sharing this information with families, you may need to:

1. **Find out.** Whether most families have enough to eat throughout the year. If not, why not. Whether most people eat a variety of foods. If not, why not. What staple and other foods are produced and eaten locally. What the blocks to increasing food production are. Whether much food is lost during on-farm storage. Who can help farmers and other food producers to produce more foods and improve storage. Which foods people buy. Which foods are good buys. What the availability of iodized salt is. What the blocks to people buying more or better food are. Whether people gather wild foods. If so, which ones. What other good wild foods could be gathered.
2. **Prioritize.** Decide which information is *most important* to share with groups or individual families.
3. **Decide whom to reach.** For example: women and men who produce, store, preserve or buy the family food, especially those from food-poor families.
4. **Choose communication methods.** For example: discussions and demonstrations with community and farmers' groups and at schools and youth clubs.

Examples of questions to start a discussion

(choose only one or two questions that deal with the information families need most)

Are there some local families who do not have enough to eat? Can we help them?

How can we produce more food? Can we produce more, different foods?

How do we store food on the farm? How can we improve our stores?

How can we budget better for food? Which local foods are best value for money? Which foods are poor buys?

Why is iodized salt a must buy?

Which wild foods do we eat? Are there others we can eat?

TOPIC 3 MAKING GOOD FAMILY MEALS

NUTRITION NOTES

Healthy, balanced diets

BOX 9 • A BALANCED DIET

A balanced diet provides the correct amounts of food energy and nutrients needed during the day to cover the dietary requirements of the person eating it. A balanced diet must be composed of a variety of different foods from different food groups so that it contains all the many macronutrients and micronutrients the person needs (see Topic 1).

In Figure 5 (page 34), the size of the various food group circles indicates the approximate, recommended (or tolerable, as in the case of sugars) amounts of each of these food groups in a healthy diet.

Topic 1 listed the main groups of foods. Topic 3 shows how to combine foods to make healthy, balanced diets. It discusses how to share meals so all of the family members get enough to cover their dietary needs.

A healthy, balanced diet contains a variety of foods

The meals and snacks a family eats during the day should provide:

- a combination of different foods. Figure 5 on page 34 shows the wide variety of foods needed to provide all the many different nutrients we require;
- enough of each nutrient to satisfy the energy and nutrient needs of each family member.

A good meal should contain:

- a staple food. Look at the list of staple foods in Topic 1, page 19, and see if it contains the local staple foods. Add them if necessary;

- other foods that may be made into a sauce, stew or relish. These should include:
  - legumes and/or foods from animals;
  - at least one vegetable;
  - some fat or oil (but not too much) to increase the energy and improve taste. Most of the fat or oil should be from foods containing unsaturated fatty acids (see Box 4 in Topic 1).

It is good to eat fruits with a meal (or as a snack) and to drink plenty of water during the day. Avoid drinking tea or coffee until 1–2 hours after a meal (when food will have left the stomach) as these reduce the absorption of iron from food.

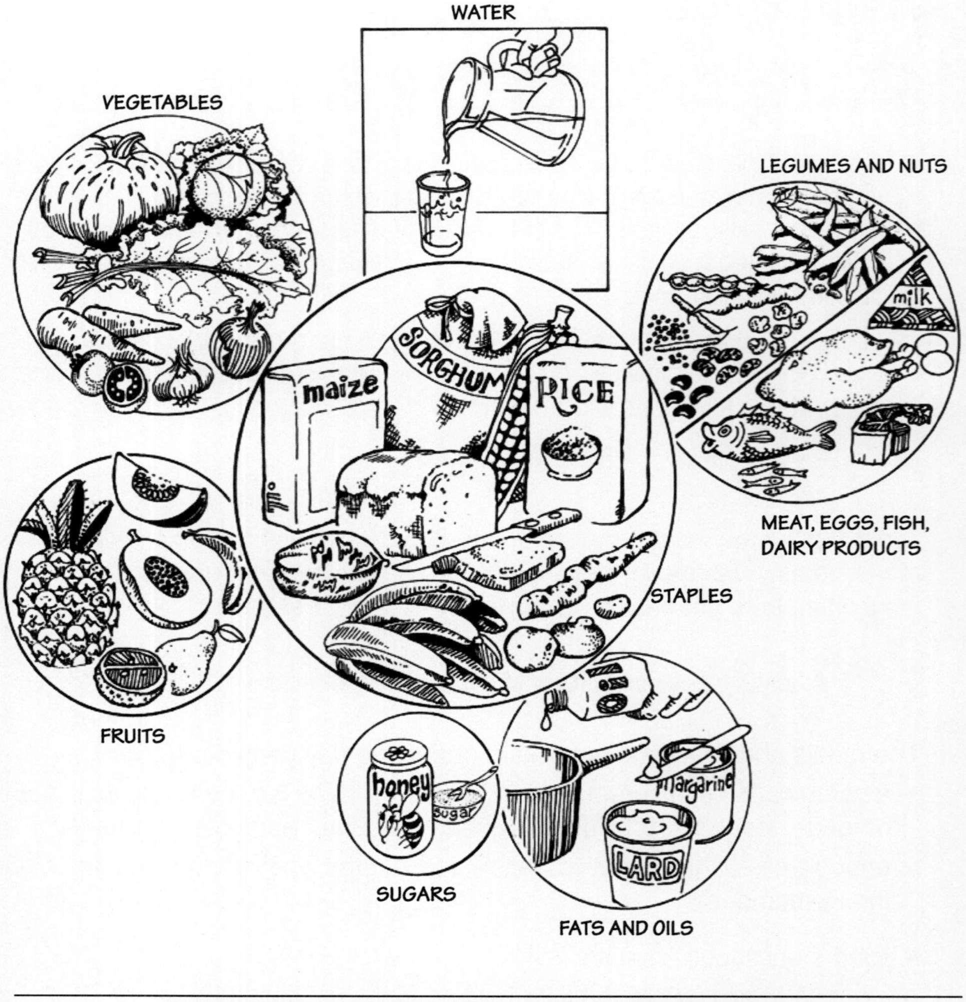

Figure 5. Use a variety of foods to make healthy meals

How to increase variety

Eat a variety of vegetables and fruits to get plenty of micronutrients and fibre

Encourage families to use:

- several groups of foods at each meal;
- different vegetables and fruits at different meals because different vegetables and fruits contain varying amounts of the different micronutrients;
- meat, poultry, offal and fish daily if possible because these foods are the best sources of iron and zinc (which are often lacking in diets, especially the diets of young children and women).

Snacks

Snacks are foods eaten between meals. Below are examples of foods that make good snacks, particularly when more than one food is eaten.

Circle the ones available in your area and add other local nutritious snacks to the list.

SNACKS

FRESH MILK, SOURED MILK, YOGHURT, CHEESE

ROASTED GROUNDNUTS, SOYBEANS, MELON SEEDS AND OTHER OILSEEDS

EGGS

FRIED FISH

BREAD, PARTICULARLY WHEN EATEN WITH MARGARINE OR GROUNDNUT PASTE/PEANUT BUTTER

BOILED OR ROASTED MAIZE COBS

BOILED OR FRIED CASSAVA, PLANTAIN, YAM, SWEET POTATOES AND POTATOES

CHAPATIS, BEAN CAKES

BANANAS, AVOCADOS, TOMATOES, MANGOES, ORANGES AND OTHER FRUITS (INCLUDING DRIED FRUITS)

YOUNG COCONUT FLESH

DATES

Eating snacks like these is a good way of improving a diet which may lack food energy and nutrients. However, people should also know that eating often throughout the day increases the risk of tooth decay, particularly where oral hygiene is poor. This is particularly true for sweet (sugary) foods that stick to the teeth.

Eating away from home

Many people buy meals and snacks from vendors or eat in bars, restaurants or hotels; some students receive snacks or a meal at school. While eating out can be a special treat, eating outside the home too often can mean getting too many calories, fat and salt and not enough fruits, vegetables and essential nutrients. The nutrient composition of food eaten away is often not of the same quality as food prepared at home.

Encourage people to choose meals and snacks that provide a mixture of nutrients, especially if they regularly eat away from home. Warn them not to eat too many fried street foods as these may contain unhealthy fats and increase the risk of overweight. Encourage people who make and sell meals and snacks to prepare good-quality foods in a hygienic way (see Topic 4).

Sharing meals

Share family meals according to energy and nutrient needs

Look at Appendix 2, Table 4. It compares the energy and nutrient needs of children, women and men of average size. If we change these nutrient needs into food needs, we find that families should:

- share staple foods and legumes according to energy needs (see Figure 6). Children aged 1–3 years need about a third of the amount needed by men. Note that energy needs increase greatly during puberty and adolescence, especially for boys, and during pregnancy and breastfeeding;
- share vegetables and fruits almost equally among all family members but make sure pregnant and breastfeeding women have bigger shares;
- give bigger shares of iron-rich foods (meat, offal, poultry and fish) to older girls and women, especially when they are pregnant. Young children are often anaemic and need a fair share of these foods too;

- make sure young children get plenty of fat-rich foods, such as milk, groundnuts, oils and fats that give them enough energy even though they eat smaller amounts of foods. Sugar, jaggery and honey are also ways to increase the energy content, and they can be added to porridge and other foods in small quantities.
- give young children their own bowl or plate. This allows them to get their share of the food if the rest of the family members are all eating from the same bowl.

Women need more iron-rich foods than men

The food needs of different family members are discussed in more detail in Topics 5–10.

Figure 6. Sharing meals according to need

Preparing and cooking good meals

Cook vegetables quickly to preserve nutrients

The way we store, prepare and cook our food affects the nutrients in it. For example, some vitamin C and folate are lost during cooking. Ways to reduce nutrient losses are:

- buying or picking vegetables and fruits on the day you use them and storing them in a cool place;
- cleaning and cutting vegetables and fresh starchy roots immediately before cooking;
- cooking vegetables in little water or with a stew until just tender; other cooking methods for vegetables that preserve nutrients are stir frying (i.e. frying very quickly over high heat), or sautéing (i.e. cooking in fat or oil in a pan or on a griddle);
- eating food soon after cooking.

We absorb the vitamin A in plant foods better when the food is cooked (but not overcooked) and eaten with fat.

Families may cook and eat less often if fuel is scarce or expensive. Ways to save fuel include:

- using fuel-efficient stoves and cooking methods;
- using dry firewood;
- soaking legumes for several hours;
- cutting food into small pieces just before cooking;
- putting a tight-fitting lid on the pot;
- cleaning soot off pans;
- putting out fires promptly.

Topic 4 explains how to prepare foods in a hygienic way.

Enjoying meals

We all enjoy our meals if they are tasty and we eat them in a comfortable happy environment. Mealtime can be a time when families talk together, entertain guests and teach young children good habits and customs. It is a time when parents can give children loving attention as they encourage them to eat.

SHARING THIS INFORMATION

Before sharing this information with families, you may need to:

1. **Find out.** What foods are eaten at different meals. What types of snacks are eaten. Which foods need promoting. What the different foods eaten by different family members are. How food is shared. What foods are eaten away from home. What the cooking facilities are. Whether people have enough different foods to make healthy meals. If not, why not. What local recipes are used.
2. **Prioritize.** Decide which information is *most important* to share with groups or individual families.
3. **Decide whom to reach.** For example: people who prepare family meals, food vendors, cooks, and school-age children.
4. **Choose communication methods.** For example: discussions and demonstrations of meal planning with women's and other community groups and at schools; leaflets with recipes.

Examples of questions to start a discussion

(choose only a few questions that deal with the information families need most)

Which combinations of local foods make good meals? How many of us make these sorts of meals regularly? If not, why not?

What can we do to improve our family meals? Which foods should we add or use more often?

Could we use more meat, offal, poultry or fish?

Could we use more different kinds of fruits and vegetables?

How do we share meals among the family? Do young children and women get their fair share? Demonstrate how a local meal should be shared. Do we share meals like this? If not, why not?

TOPIC 4 KEEPING FOOD SAFE AND CLEAN

NUTRITION NOTES

Why foods and drinks must be safe and clean

It is important that the food we eat and the water we drink is clean and safe. So it is essential to prepare meals in a safe, hygienic way. If germs (such as harmful micro-organisms and parasites) get into our foods and drinks, they may give us food poisoning (resulting, for example, in diarrhoea or vomiting). The people most likely to become sick are young children and people who are already ill, particularly people living with HIV/AIDS.

We can prevent most food poisoning by following a few basic and simple rules of hygiene that aim to:

- prevent germs from reaching foods and drinks. Many germs come from human or animal faeces. Germs can reach:
  - food via dirty hands, flies and other insects, mice and other animals and dirty utensils;
  - water supplies if they are not protected from faeces.
- prevent germs from multiplying in foods and reaching dangerous levels. Germs breed fastest in food that is warm and wet (e.g. porridge), especially if it contains sugar or animal protein, such as milk.

To help families have clean, safe foods and drinks:

- find out about disposal of faeces, hand washing practices, the source and storage of water and ways in which food is prepared. This helps you identify ways in which germs may be reaching food and water, and foods in which germs may be breeding;
- suggest practical ways to improve water and food hygiene. Some of the suggestions listed below may be relevant and useful. But remember not to overburden families with too much advice.

Personal hygiene

Wash hands after contact with faeces

Advise people to:

- wash hands with clean water and soap (or ashes):
  - after going to the toilet, cleaning a baby's bottom or cleaning clothes, dirty bed linen or surfaces contaminated with faeces. It is most important to wash hands after contact with faeces;
  - before and after preparing food and eating;
  - before feeding a child or sick person (make sure they wash their hands too).
- dry hands by:
  - shaking and rubbing them together *or;*
  - using a clean cloth that is kept only for this purpose.
- keep fingernails short and clean;

Figure 7. Washing hands helps prevent disease

- avoid coughing or spitting near food or water;
- cover any wounds on hands to prevent contamination of food during its preparation;
- use a latrine and keep it clean and free of flies;
- teach small children to use a potty. Put children's faeces in the latrine;
- clean up faeces from animals.

Dispose of faeces safely

Clean and safe water

Use water that comes from a safe source or is boiled before drinking

Advise families to:

- use safe water, such as treated pipe water, or water from a protected source, such as a borehole or protected well. If the water is not safe, it should be boiled (rapidly for one minute) before it is drunk or used in uncooked foods (e.g. fruit juices);
- use clean covered containers to collect and store water.

Buying and storing food

Cover foods to keep them clean and safe

Advise families to:

- buy fresh foods, such as meat or fish, on the day they will eat them. Look for the signs of poor-quality food (see Topic 2, page 29);
- cover raw and cooked foods to protect them from insects, rodents and dust;
- store fresh food (especially foods from animals) and cooked foods in a cool place, or a refrigerator if available;
- keep dry foods such as flours and legumes in a dry, cool place protected from insects, rodents and other pests;
- avoid storing leftover foods for more than a few hours (unless in a refrigerator). Always store them covered and reheat them thoroughly until hot and steaming (bring liquid food to a rolling boil).

Preparing food

Prevent raw meat, offal, poultry and fish from touching other foods

Advise people preparing food to:

- keep food preparation surfaces clean. Use clean, carefully washed dishes and utensils to store, prepare, serve and eat food;
- prepare food on a table where there is less dust;
- wash vegetables and fruits with clean water. Peel if possible;
- prevent raw meat, offal, poultry and fish from touching other foods, as these animal foods often contain germs. Wash surfaces touched by these raw foods with hot water and soap;
- cook meat, offal, poultry and fish well. Meat should have no red juices;
- boil eggs so they are hard. Do not eat raw or cracked eggs;
- boil milk unless it is from a safe source. Soured milk may be safer than fresh milk.

Hygiene around the home

Advise families to:

- keep the surroundings of the home free from animal faeces and other rubbish;
- keep rubbish in a covered bin and empty it regularly so it does not attract flies;
- make compost for the garden with suitable waste food, garden rubbish and animal faeces. Composting destroys germs in faeces.

Toxins and chemicals

Food and water is unsafe if it contains toxins or dangerous chemicals. A toxin called "aflatoxin" is made by a mould that grows on cereals and legumes. Eating aflatoxin can make us seriously ill. Advise families to prevent moulds from growing by drying crops thoroughly and storing them in a dry place. Warn people not to eat mouldy foods or give them to animals. They can add them to compost.

Pesticides and other harmful agricultural chemicals may get into food or water and cause poisoning if:

- the chemical is not used in the recommended way;
- the empty containers are used for food or water.

Advise people to:
- follow carefully the instructions for using chemicals;
- be strict about keeping chemicals away from children;
- never put food or water into containers that have been used for chemicals;
- wash hands after using chemicals, and wash any foods (e.g. fruit) that have been sprayed with them.

SHARING THIS INFORMATION

Before sharing this information with families, you may need to:

1. **Find out.** What the sources and quality of household water supplies are. What the local hygiene practices are, particularly those related to washing hands and getting rid of adults' and children's faeces. How food is stored and prepared. What the principal unhygienic food and personal practices in the area are. What people know about keeping food and water safe and clean. How agricultural chemicals are used and how they are handled.
2. **Prioritize.** Decide which information is *most important* to share with groups or individual families.
3. **Decide whom to reach.** For example: women and others who prepare food or fetch water.
4. **Choose communication methods.** For example: health talks, discussions and demonstrations (e.g. washing hands), with community groups and at clinics and homes.

Examples of questions to start a discussion

(choose only one or two questions that deal with the information families need most)

Why is it important to prepare food in a hygienic way?

When should we wash our hands? How should we wash and dry our hands?

Why is it important to get rid of faeces from adults and children safely? How can we do this?

Is the local water supply safe to drink? If not, what should we do?

Is the local milk safe to drink? If not, what should we do?

Why should we prevent raw meat, poultry and fish from touching other foods? How can we do this?

How should we store different types of food (e.g. vegetables, meat, cooked foods)?

How should we deal with waste from food?

What should we do with mouldy food?

TOPIC 5 FOOD AND CARE FOR WOMEN

NUTRITION NOTES

Why women should eat well

Well-nourished mothers are likely to have healthy babies

Girls and women need to eat well throughout their lives but particularly when they are planning a baby, are pregnant or breastfeeding. If they eat healthy, balanced diets they are likely to:

- stay active and well;
- produce healthy babies and breastfeed successfully.

Low birthweight babies are more likely to grow and develop more slowly than healthy babies

A woman is at risk of complications and a difficult labour if she is already undernourished when she becomes pregnant, or is undernourished during pregnancy, and her baby is likely to have a low birthweight (i.e. <2 500 g). Low birthweight babies are at greater risk than healthy newborns of:

- growing and developing slower;
- contracting an infection and of dying. The lower the birthweight the greater the risk of death;
- having low body stores of micronutrients that may result in disorders, such as anaemia, and vitamin A and zinc deficiencies;
- developing heart disease, high blood pressure, obesity and diabetes when adult.

Other causes of low birthweight are prematurity, malaria or other infections in the mother, or the mother's smoking or abusing drugs during pregnancy.

TOPIC 5

Feeding women and girls of reproductive age

Women and older girls need plenty of iron-rich foods

Look at Appendix 2, Table 4. It compares the daily energy and nutrient needs of average-sized women and men. Women of reproductive age who are not pregnant or breastfeeding have slightly lower energy and protein needs than men but they need double the amount of iron (because of menstruation). Compared to men's diets, the diets of women should provide:

- slightly smaller amounts of staples, legumes and fats;
- at least the same amounts of vegetables and fruits;
- more iron-rich foods (meat, offal, poultry and fish).

Nutrient needs increase during pregnancy and breastfeeding

Women's needs for energy and most nutrients increase during pregnancy and breastfeeding. Iron needs during pregnancy are so high that it is usually advisable to give iron supplements, such as iron/folic acid tablets (see Topic 11, page 91).

Make sure that women and their relatives know the following.

- All girls and women of reproductive age should:
  - eat a healthy, balanced diet (see Topic 3) that contains plenty of iron-rich foods;
  - have plenty of clean, safe drinks;
  - eat iodized salt. Women who lack iodine when they become pregnant are at greater risk of having a baby who is physically and mentally damaged (see Section C, page 9).
- Pregnant and breastfeeding women and girls need extra food (see Appendix 2, Table 4).
  - When pregnant they need about 280 extra kcal/day, more protein, zinc, vitamin A, vitamin C and folate, and much more iron (i.e. the equivalent of an extra nutritious snack each day; see Topic 3, page 35, for examples). It is particularly important for women to eat well and be well nourished *throughout* their pregnancy, including the first trimester, so that the babies' bodies and brains develop properly. Women should gain about 1 kg a month in the second and third trimester of pregnancy.
  - When breastfeeding they need about 450 extra kcal/day and much more protein, zinc, vitamin A, vitamin C and folate (i.e. the equivalent of an extra small meal each day). You can suggest that women eat more at each meal or eat more frequently – perhaps having more snacks during the day.

- Women should eat well between pregnancies so they rebuild their bodies' stores of nutrients.

A woman who is overweight or obese when she becomes pregnant should eat healthy meals but not 'diet'. Advise her how to lose weight if she is still overweight after breastfeeding (see Box 19 in Topic 11, page 94).

At certain times some women may need micronutrient supplements in addition to good meals. For example, most women need iron/folic acid tablets during pregnancy. A good diet should provide enough of the other micronutrients, including vitamin A. However, in situations where vitamin A is likely to be deficient, women should receive vitamin A supplements *as soon after delivery as possible and not more than six weeks later.* This provides a store for use during breastfeeding. Do not give high doses of vitamin A to any woman who could be pregnant as they can harm her unborn baby.

Figure 8. Women need extra food when they are pregnant or breastfeeding

In some places many women are HIV+. Make sure these women know:

- the risks of passing the virus to their unborn or breastfeeding babies and how to minimize these risks (see Topic 6, page 55);
- that good feeding will help them stay healthy longer (see Topic 10, page 85).

Another way to help women and their unborn babies

Spacing births can improve the health of women and babies

You can help to improve the health of women and prevent their babies from having low birthweights by encouraging family planning. Advise parents to:

- wait at least two to three years between pregnancies;
- not have a baby when the woman is too young (e.g. under 18 years) or too old;
- wait at least six months between ending breastfeeding and becoming pregnant again. This gives time for women to 'fill up' their body stores of fat, iron and other nutrients and become strong again.

Exclusive breastfeeding (see Topic 6) is one contraceptive method (although not a totally secure one). A woman is unlikely to become pregnant if:

- she has not restarted her menstrual periods *and*;
- the baby is less than six months old *and*;
- the baby breastfeeds exclusively (has nothing else to eat or drink or suck).

Dangers of adolescent pregnancy

Adolescent mothers are likely to be undernourished and have low birthweight babies

Adolescent pregnancy is a nutritional as well as a social problem in many places. Adolescent mothers are likely to be undernourished and to have undernourished babies because:

- their bodies are still developing so their nutrient needs during pregnancy are especially high. They are more likely to die during pregnancy and childbirth than older women;
- some girls are frightened to admit that they are pregnant, so they delay getting antenatal care. Some girls are forced to leave school or home and to support themselves, often by prostitution.

Warn adolescent girls of the dangers of becoming pregnant, tell them about the different methods of contraception, and monitor and counsel them sympathetically if they do get pregnant.

SHARING THIS INFORMATION

Before sharing this information with families, you may need to:

1. **Find out.** What women eat. What pregnant women eat. What breastfeeding women eat. What the food customs and taboos for menstruating, pregnant and breastfeeding women are. What types of malnutrition there are among women, especially pregnant and breastfeeding women. Whether adolescent pregnancy is a problem. Whether adolescent pregnant girls are undernourished. How many babies have low birthweights. What the causes of low birthweight are. What people believe are the causes of low birthweight. What the blocks to women having better diets are. Whether anaemia and/or vitamin A deficiency disorders are problems in the area.
2. **Prioritize.** Decide which information is *most important* to share with women and their families.
3. **Decide whom to reach.** For example: women and adolescent girls; women's partners and other relatives; relatives of adolescent girls.
4. **Choose communication methods.** For example: discussions, handouts, demonstrations of good foods for women, quizzes, plays/drama and songs.

Examples of questions to start a discussion

(choose only a few questions that deal with the information women and their partners need most)

Why do women and girls need good food all the time?

Do pregnant women need extra food? Why? Which foods are good for pregnant women? Do women need to improve their diets when they are pregnant? How can they do this?

Are there customs and taboos that prevent some women from eating nutrient-rich foods (e.g. eggs or fish)?

Do breastfeeding women have special food needs? What are they? Do they need to improve their diets when they are breastfeeding? How can they do this?

Do some babies have low birthweights? Does this matter? What can we do to improve the birthweights of the babies?

Are adolescent pregnancies a problem in the area? Why are adolescent mothers at risk of having low birthweight babies? How can we help these girls?

TOPIC 6 FEEDING BABIES AGED 0-6 MONTHS

NUTRITION NOTES

How you help a mother to feed her young baby depends on whether the mother is HIV– (negative), of unknown HIV status or HIV+ (positive). Much research is presently being done on the feeding of babies whose mothers are HIV+. The advice given in this topic is what nutritionists currently (in 2004) recommend (see WHO/UNICEF/UNFPA/UNAIDS. 2003. *HIV and infant feeding* listed in Appendix 3).

BOX 10 • EXCLUSIVE BREASTFEEDING

Exclusive breastfeeding means an infant receives *only* breastmilk from the mother or a wet nurse, or expressed breastmilk, and no other liquids or solids except drops or syrups consisting of vitamins, mineral supplements or medicines.

If the mother is HIV– or of unknown HIV status

Most babies should breastfeed exclusively for six months

Advise the mother to exclusively breastfeed until the baby is six months (180 days) old.

- Breastmilk contains all the nutrients a full-term baby needs for the first six months of life. It provides enough water even in hot weather and is the safest source of water.
- Exclusive breastfeeding reduces the risk of diarrhoea and other infections. Giving any other food or drink increases the risk of diarrhoea.
- Exclusive breastfeeding means the mother is unlikely to become pregnant.

Breastmilk provides all the food and water young babies need

Ways to encourage exclusive breastfeeding include:

- helping the baby to start suckling within one hour of birth – the mother and baby should be in skin contact immediately after birth;
- if necessary, explaining why colostrum is an essential food for newborn babies. Colostrum contains high levels of vitamin A and anti-infective factors that protect newborns from disease. Giving colostrum is like giving a first immunization. If a family has a wrong belief about colostrum (e.g. it is dirty), help them to understand it is safe, and is the perfect food for their new baby;

Figure 9. Suckling in the correct position

Baby's body is turned towards mother, the chin touches mother's breast, the mouth is wide open and both lips are turned outwards. More areola is above than below baby's mouth. The baby takes slow deep sucks and you can hear the baby swallowing.

- checking that the baby is suckling correctly (see Figure 9);
- if necessary, explaining why families should not give baby any other food or drink (even traditional drinks);
- advising the mother to feed 'on demand' (when the baby wants to feed) at least 8–10 times over 24 hours, and let the baby suckle for as long as he or she wants day and night;
- dealing with breastfeeding problems (e.g. sore nipples, engorged breasts, thrush in baby's mouth) promptly;
- teaching the mother how to express and store her milk if she is away from her baby for more than three hours;
- referring the mother to a local breastfeeding support group if there is one.

Colostrum is the best and safest food for newborns

Also advise families that breastfeeding mothers need:

- extra food (the equivalent of one extra small meal a day). They especially need more meat, poultry, offal and fish, and more vegetables and fruits;
- enough drink so they are not thirsty;
- more rest if possible.

Make sure mothers know that HIV can be passed to their babies through breastmilk and how to avoid that their babies become infected.

If the mother is HIV+

Explain the risks and benefits of breastfeeding and replacement feeding to HIV+ mothers and their partners before the baby is born

While the mother is still pregnant:

- explain to her the risk of the virus being passed to her baby through breastmilk;
- explain and discuss the risks and benefits of exclusive breastfeeding and of replacement feeding, and the risks of feeding breastmilk with other foods (see Box 11, page 56).

Replacement feeding means feeding a child who is not receiving breastmilk with a diet that provides all the nutrients the child needs. During the first six months this should be a suitable breastmilk substitute, such as commercial or home-made formula.

BOX 11 • RISKS AND BENEFITS OF DIFFERENT WAYS OF INFANT FEEDING

Exclusive breastfeeding

- It gives immunity from other infections, is the best source of nutrients and safe water, reduces the risk of pregnancy and prevents the possible stigma of not breastfeeding.
- There is a risk of passing HIV to the baby but this is lower if:

  the mother gives no other food or drink;

  the mother does not have cracked nipples or mastitis, or is not clinically ill with AIDS;

  the baby does not have sores or thrush in the mouth.

Replacement feeding

- There is no risk of passing HIV to the baby.
- There is a high risk of diarrhoea and other infections if the family lacks the resources to buy and prepare other milk feeds safely.
- There is a risk that the caregiver will prepare the feed incorrectly (e.g. over-dilute it) so that the child becomes malnourished.
- There is the possibility of stigma and of others knowing the mother's HIV status.

Replacement feeds should only be given where they are acceptable, feasible, affordable, sustainable and safe.

Feeding both breastmilk and breastmilk substitutes

- There is a higher risk of passing HIV to the baby than with exclusive breast-feeding.
- There is a risk of other infections and malnutrition if breastmilk substitutes are not prepared safely and correctly.

When a HIV+ mother has decided how to feed her baby, give her support and advice. If the mother agrees, try to talk with relatives (e.g. her husband, partner and/or mother) so they can also support and help her.

If the mother decides to breastfeed:

- strongly advise her to start exclusive breastfeeding immediately after birth, and not to give any other food or drink. Advise her to exclusively breastfeed for the first few months and up to six months. When she wants to stop breast-feeding, she should do this when the family is able to give suitable replacement feeds;

- take time to explain the risks of feeding breastmilk with other foods;
- counsel her on how to exclusively breastfeed (see above);
- advise her to immediately seek health care if she has cracked nipples, engorged breasts or if her baby has sores or thrush in the mouth;
- counsel, in advance, on how to stop breastfeeding as this should be done at an earlier age and over a shorter period than usual, and the mother needs to plan for this change (see Box 12);
- weigh the baby at least monthly to monitor his or her growth.

BOX 12 • STOPPING EXCLUSIVE BREASTFEEDING FOR HIV+ MOTHERS

HIV+ mothers should stop breastfeeding over a *shorter period than usual* (i.e. the change-over period from exclusive breastfeeding to replacement feeding should last only about two weeks or less). This is because the baby is at higher risk of HIV infection during the change-over period.

However, ceasing breastfeeding over a short period increases the risk of difficulties such as mastitis and breast abscesses, and objections from families – and the babies may become distressed and lose their appetites.

To help mothers and babies during the change-over period, health workers can:

- show a mother how to express her breastmilk and then heat-treat it (heat-treating destroys the HIV virus). This reduces the risk of engorgement for the mother and allows the baby to continue receiving breastmilk while becoming used to the tastes of replacement feeds and to cup feeding. To heat-treat breastmilk, put the milk in a small pot, heat until the milk boils and then put the pot into a container of cold water so the milk cools quickly;
- advise a mother (and her relatives if possible) on suitable replacement feeds and how to prepare them. Babies aged less than six months should receive only breastmilk substitutes (home-made or commercial infant formula) or heat-treated breastmilk. After that they should also have complementary feeds (see Topic 7);
- tell a mother to give extra attention and love to her baby and to give replacement feeds or expressed heat-treated breastmilk herself;
- advise a mother to seek health care immediately if she has any signs of mastitis and/or sore nipples;
- if appropriate, explain to relatives the reasons for ceasing breastfeeding earlier than usual.

If the mother decides not to breastfeed:

- advise the mother (or other caregiver) not to give any breastmilk (unless expressed and heat-treated). Emphasize the risks of giving both breastmilk and other foods;
- check that the family has the resources and skills for making and giving replacement feeds;
- show the mother how to prepare the feeds and how to feed with a cup. Emphasize the need for good hygiene and for diluting the milk correctly. Explain the risks of using a bottle (e.g. they are difficult to clean and so increase the risk of diarrhoea);
- watch the mother prepare and give a feed and correct any mistakes. Try to do this in her own home using her own equipment;
- encourage the mother to feed the baby herself and to cuddle him or her as often as possible;
- if appropriate, talk with the mother's relatives (e.g. her partner or mother) and explain what they can do to support and help her;
- tell the family to take the baby quickly to a health worker if there are any feeding or health problems.

Monitoring baby's weight

- Babies aged 0–6 months should be weighed at least monthly. Plot the weights on a growth chart and make sure the mother or caregiver understands the growth curve (see Topic 11, page 89). This is especially important for children whose mothers are HIV+.
- Give any necessary advice and support on feeding and care (see Topic 11). Topic 7 explains when to start complementary foods.
- Give vitamin A supplements according to national protocols.

SHARING THIS INFORMATION

Before sharing this information with families, you may need to:

1. **Find out.** How local babies aged 0–6 months are fed. Whether mothers exclusively breastfeed, and if so, for how long. If not, which other foods, water or other drinks are given. What the blocks to exclusive breastfeeding for six months are. How women who are HIV+ feed their babies. What their knowledge of the risks and benefits of different feeding methods is. Who decides how babies are fed. What advice and resources are needed by mothers who decide not to breastfeed. Which breastmilk substitutes are available locally and what their costs are. What breastfeeding women do if they have breastfeeding problems, such as sore nipples or engorged breasts, or if their babies have thrush.
2. **Prioritize.** Decide which information is *most important* to share. This may depend on whether you are communicating with groups of mothers or parents, with individual HIV+ mothers, with mothers who are HIV– or whose status is unknown, or with traditional midwives.
3. **Decide whom to reach.** For example: mothers, other caregivers and, if appropriate, their partners and other relatives; traditional midwives.
4. **Choose communication methods.** For example: individual counselling and group discussions at antenatal and postnatal clinics, in maternity wards and at young child clinics; demonstrations of suckling position, replacement feeding and heat-treating expressed breastmilk.

Examples of questions to start a discussion

(choose only a few questions that deal with the information families need most)

What is exclusive breastfeeding? Why do we recommend exclusive breastfeeding?

Why is colostrum an excellent food for newborns? Do we give colostrum to our babies? If not, why not?

Do breastfed babies need extra water?

What foods or drinks other than breastmilk do we sometimes give young babies? Why? Could we stop doing this?

What should women who have sore nipples or engorged breasts do?

Do breastfeeding women need extra food? Which foods are good for breastfeeding women?

Discuss the feeding of babies of HIV+ mothers only if a group wants to. Do this in a sensitive way. Otherwise counsel mothers individually.

Can the virus be passed to a baby through breastmilk? Explain that the risk may be less if a baby is exclusively breastfed.

What are the dangers of replacement feeding? Explain the risks and benefits of both exclusive breastfeeding and replacement feeding. Explain the risks of giving both breastmilk and breastmilk substitutes.

How can breastmilk be made safe during the time that a mother is changing from exclusive breastfeeding to replacement feeds? Explain why and how to express and heat-treat breastmilk.

If replacement feeding occurs in the area

Which breastmilk substitutes are available and used here? How much do they cost? Do mothers know how to prepare them in a safe and hygienic way? Are they culturally acceptable?

Why is it dangerous to feed with a bottle? Do women know how to feed with a cup?

Demonstrate preparing and giving a feed using a breastmilk substitute that local families can afford.

TOPIC 7 FEEDING YOUNG CHILDREN AGED OVER SIX MONTHS

NUTRITION NOTES

BOX 13 • COMPLEMENTARY FEEDING

Complementary feeding means giving other foods in addition to breastmilk (or breastmilk substitutes).

Previously, the term 'weaning' was used, but there was some confusion about its meaning. Some people thought that weaning meant 'stopping breastfeeding'; others thought it meant 'the period during which the child changed from having only breastmilk to only family foods'. 'Weaning foods' could mean foods given when the child stopped breastfeeding or during the change from breastmilk to family foods.

Always using the term 'complementary feeding' (also when translated into a local language), should avoid this confusion and ensure that the recommended feeding for children aged over six months is: continued breastfeeding plus complementary foods.

When to start complementary feeding

Start complementary foods when a baby is six months old

Topic 6 explains why most babies need only breastmilk for the first six months of life. Most babies should start complementary foods when they are six months old because at this age:

- breastmilk alone cannot supply all the nutrients needed for growth;
- children are able to eat and digest other foods.

Most children should breastfeed for two years and, if possible, beyond

Topic 7 discusses what foods to give children aged over six months. Breastfeeding on demand should continue until a child is 2 years old or beyond (unless the mother is HIV+: see below and Topic 6). As children grow, increasing amounts of complementary foods are needed to fill the gap between the nutrients supplied by breastmilk and children's nutrient needs.

What to give and when

Advise parents to start by giving 1–2 teaspoons of semisolid food, for example porridge or mashed potato, and to add other foods to make good complementary meals (see below). By the age of eight months, babies also like 'finger foods', foods they can hold themselves, such as a chapati or banana. By the age of 1 year, most children can eat suitable family meals and snacks.

Prepare complementary foods hygienically – Keep clean

Good complementary foods:

- are rich in energy, protein and micronutrients, especially iron, and are not watery (i.e. thick not thin porridges);
- are easy to eat and digest;
- are hygienically prepared and fed (see Topic 4);
- contain no bones or hard pieces that might cause choking;
- are not too spicy or salty. Too much salt is bad for children.

Use a variety of foods for children's meals

Advise parents to prepare meals that provide:

- a variety of foods (see Topic 3);
- some fat-rich foods to increase the energy content;
- fresh fruits and vegetables, especially ones rich in vitamins A and C;
- eggs, milk foods and iron-rich animal foods (meat, offal, poultry, fish, as appropriate) daily or as often as possible.

Young children also need snacks. Here are examples of good snacks for young children.

Circle the snacks used locally and add others to the list.

SNACKS FOR YOUNG CHILDREN
FRUITS SUCH AS MANGO, PAWPAW, BANANA, AVOCADO
BOILED EGG
BOILED, PASTEURIZED OR SOURED ANIMAL MILK
CHAPATI OR BREAD WITH GROUNDNUT PASTE/PEANUT BUTTER OR MARGARINE OR DIPPED IN MILK
SMALL PIECES OF BOILED OR FRIED CASSAVA, PLANTAIN OR YAM
SWEET POTATOES (ORANGE COLOURED)

Porridges made with germinated or fermented flours. Young children need foods rich in energy and nutrients because their stomachs are small and they cannot eat large amounts at each meal. Porridge is the most common food for young children, but its energy and nutrient content is often too low to meet their needs fully. This is due to the high starch content of staple foods, such as maize, millet, sorghum, cassava and yams. During cooking, these flours absorb much water, which makes them bulky and thick. If water is added to make the porridge less thick and easier for young children to eat, its energy and nutrient content is further decreased. Children would need to eat large quantities of such diluted porridge in order to meet their energy and nutrient needs, but because of their small stomachs it is difficult for young children to consume large quantities.

Ways to make porridges more energy and nutrient-rich, *and* easy to eat are by:

- adding energy-rich (e.g. oil, butter/ghee) and nutrient-rich foods (such as flours of groundnut and other legumes, or sunflower seed) to the porridge;
- making porridges with germinated or fermented cereal flours (see Box 14, page 64).

Porridges made with germinated or fermented cereal flours do not thicken as much as ordinary porridges. They can be made with less water and so contain more energy and nutrients in a smaller volume. Other advantages of these flours are the following:

- the iron is better absorbed than from plain (non-germinated and non-fermented) flours;
- porridge made from fermented flour is easier to digest and safer because germs cannot grow easily in fermented porridge.

BOX 14 • MAKING GERMINATED FLOUR AND PORRIDGES WITH GERMINATED OR FERMENTED CEREAL FLOURS

How to make germinated flour

1. Sort and clean the cereal grains.
2. Soak them for 1 day.
3. Drain and place in a sack or other covered container.
4. Store in dark, warm place for 2-3 days until the grains sprout.
5. Dry the sprouted grains in the sun.
6. Grind and sieve the flour.

How to make fermented porridge

There are many ways to make this. One way is to:

1. Grind cereal grains into flour.
2. Soak the flour in water (about 3 cups flour to 7 cups water).
3. Leave to ferment for 2–3 days.
4. Cook into porridge.

Sometimes flours from starchy roots (e.g. cassava) are made into fermented porridges.

Recipes for good porridges

You can use germinated or fermented flours to make good porridges. The recipes below suggest porridges that are energy and nutrient-rich.

They have been adapted from the following source: FAO. 2001. *Improving nutrition through home gardening. A training package for preparing field workers in Africa,* and WHO. 2000. *Complementary feeding: family foods for breastfed children.*

Millet and bean porridge (serves 4)

cowpea leaves
1 cup pigeon pea flour
3 cups millet flour (germinated, fermented or plain)
10–12 cups hot water

Sort the cowpea leaves, boil for about 5 minutes, dry and pound. Mix the pigeon pea and millet flours, add cold water and mix into a smooth paste. Add hot water to the paste and cook, stirring constantly, until the porridge is ready. Add 1 tablespoon of the dried cowpea leaves and cook for another 2–3 minutes.

To prepare a smaller amount of the same porridge (for one child's portion), use:
1/4 cup pigeon pea flour
3/4 cup millet flour (germinated, fermented or plain)
1/4 tablespoon dried and pounded cowpea leaves

Zimbabwe multimix

1/2 cup maize flour (germinated, fermented or plain)
2 tablespoons bean flour
1 teaspoon chopped spinach
1/2 tablespoon vegetable oil

Mix maize and bean flour and cook to make a thick porridge. Add the spinach and vegetable oil and cook for another 2 minutes.

Cereal, groundnut, egg and spinach porridge
(for one meal for a breastfeeding child)

4-1/2 tablespoons thick cereal (e.g. maize) porridge made with germinated, fermented or plain flour
1 tablespoon groundnut paste or flour
1 egg
1 handful chopped spinach

Add the groundnuts to the porridge. Add the raw egg and spinach and cook for a few minutes.

Legume flours are useful for enriching cereal and root or tuber flours used to prepare infant feeds. Box 15 shows the step-by-step stages in the processing of cowpea, pigeon pea and soybean flours.

BOX 15 • MAKING LEGUME FLOURS

To prepare the legume flours, the legumes are cleaned and any rotten grains or unwanted materials are sorted out. The legumes are then roasted and milled or ground. The flour is sieved to remove any remaining large pieces.

How to make cowpea flour

1. Sort and wash cowpeas.
2. Roast them.
3. Peel them (optional).
4. Pound or grind them.
5. Sieve flour.

How to make pigeon pea flour

1. Sort and wash pigeon peas.
2. Soak in water for 2–3 minutes. Drain them.
3. Cover with banana leaves and leave for 6 days.
4. Roast them.
5. Mill or pound them into flour.
6. Sieve flour.

How to make soybean flour

1. Sort soybeans; do not wash them.
2. Bring water to a boil.
3. Drop beans into boiling water and boil for 10 minutes. Drain them.
4. Roast them.
5. Peel them.
6. Roast them again.
7. Mill or pound them.
8. Sieve flour.

How often to feed

Feed young children frequently

The appropriate number of feedings depends on the energy density of the local foods and the usual amounts consumed at each feeding. Young children have small stomachs, so they should eat often, with an increasing number of times as he/she grows older. For the average healthy and frequently breastfed child, complementary foods should be given as follows:

- 2–3 meals a day at ages 6–8 months;
- 3–4 meals a day at ages 9–24 months;
- with additional 1–2 good snacks (see page 63) offered each day as desired after the age of six months.

Encouraging young children to eat

Encourage young children to eat

Young children are often slow and messy eaters who are easily distracted. They eat more when their parents supervise mealtimes and actively and lovingly encourage them to eat (see Figure 10, page 68). This is especially important when children start complementary foods and until they are at least 3 years old.

Suggest that mothers, or the main caregivers:

- sit with children and encourage them to eat by talking with them and telling them how good the food is;
- make mealtimes happy times;
- feed young children with the rest of the family but give them their own plates and spoons to make sure they get, and eat, their share;
- give foods that children can hold if they want to feed themselves and tell them not to worry about messy eating – but make sure that all the food eventually gets into a child's mouth;
- mix foods together if a child picks out and eats only favourite foods;
- do not hurry children. A child may eat a bit, play a bit, and then eat again;
- make sure the child is not thirsty because thirsty children eat less, but do not fill up the child's stomach with too much drink before or during the meal;
- try to feed children as soon as they are hungry; do not wait for them to start crying for food;
- do not feed when children are tired or sleepy;
- make mealtimes interesting learning times; for example, teach the names of foods.

Sometimes even healthy children are fussy eaters. Check that the child is not sick, undernourished or unhappy and then advise families to:

- give more attention when the child eats well and less when the child is trying to gain attention by refusing food;
- play games to persuade a reluctant child to eat more;
- avoid force-feeding because this increases stress and decreases appetite even more.

For more information on complementary feeding see: WHO. 2000. *Complementary feeding: family foods for breastfed children* (listed in Appendix 3).

Figure 10. Actively encouraging a young child to eat

Children whose mothers are HIV+
(also see Topic 6, page 55)

The risk of passing HIV through breastmilk increases if a child has other foods *as well as* breastmilk. Therefore an HIV+, breastfeeding woman should *exclusively* breastfeed for a few months. When she wants to stop, she should:

- do this over a much shorter period than usual (see Box 12, page 57);
- give suitable replacement feeds. When the child is more than six months old, these feeds can be nutritious family foods, including as much food from animals as possible (i.e. milk and foods made from milk, eggs, meat, offal, poultry and fish).

Children aged over 3 years

By the age of 3 years, most children can feed themselves. But families should continue to watch and encourage children at mealtimes, especially if they are sick. Give family meals that contain a variety of different foods (see Topic 3) and are not too spicy, sugary or salty. Give three meals and 1–2 snacks a day. Where families eat from the same pot, it is a good idea to give young children their own plate or bowl so they receive their fair share of food.

SHARING THIS INFORMATION

Before sharing this information with families, you may need to:

1. **Find out.** When children usually start complementary foods. What foods are given at different ages. How foods are prepared (hygiene and consistency). How often children of different ages are fed. Who feeds the child, where and how. When and how HIV+ mothers start replacement feeding. What the blocks to the better feeding of young children are (there are likely to be several related to what children eat and how they are fed). Which blocks are the most important. Which should you try to remove first. What families say their problems are in feeding young children (e.g. mothers' time constraints).
2. **Prioritize.** Decide which information is *most important* to share with groups or individual families.

TOPIC 7

3. **Decide whom to reach.** For example: mothers and caregivers of young children; other relatives who feed children, influence feeding practices (e.g. fathers) or can help mothers; staff of child care centres and nurseries.
4. **Choose communication methods.** For example: discussions with women's groups and at young child clinics; demonstrations of complementary meals and snacks, and how to feed them.

Examples of questions to start a discussion

(choose only a few questions that deal with the information families need most)

When should most children start to eat foods in addition to breastmilk?

Which are good foods for children aged 6–8 months?

How often should we feed complementary meals to children aged 6–8 months? 9–11 months?

Do we give young children snacks? What do we give? Are there other local snacks that are good for young children?

Do you use germinated or fermented flours to make porridge? Why are porridges that are made with germinated or fermented flours good foods for young children?

Do we actively encourage young children to eat? Discuss who feeds a child, where the child is fed and what utensils are used. Demonstrate how to feed a young child.

TOPIC 8 FEEDING SCHOOL-AGE CHILDREN AND YOUTHS

NUTRITION NOTES

Why older children need good food

Food needs are high during adolescence

Like other members of the family, children of school age and youths need to eat healthy, balanced diets. It is especially important that girls eat well so that when they are women, they are well nourished and can produce healthy babies.

Appendix 2, Table 4, shows the nutrient needs of older children. Notice that:

- the needs for most nutrients increase as girls and boys reach puberty because they are growing so quickly and often gain half their final body weight during adolescence (10–18 years). Adolescent boys have especially high energy needs and that is why they are often hungry and eat large quantities of food;
- girls' needs for iron more than double when they start to menstruate. After this time and until menopause, girls and women always need much more iron than boys and men (see Topic 5, page 48);
- if adolescent girls become pregnant, they have even higher nutrient needs. These can be met by giving larger or more frequent meals and snacks, selecting foods particularly high in nutrient content, and ensuring that the diet includes a wide variety of foods. The combination of pregnancy and growth makes iron needs so high that it is usually advisable to give iron supplements.

All children, especially girls, need iron-rich diets

What happens if children are not well fed

Hungry children cannot study well

Older children who are hungry or who have poor diets are likely to:

- grow slowly;
- have little energy to play, study or do physical work;
- be anaemic and/or lack vitamin A or iodine (see Section C, page 9, and Topic 11, page 91).

Children who are hungry have short attention spans and do not do as well at school as they should.

Overweight and obesity among children and youths are becoming problems in some places, especially urban areas. Children, like adults, are at risk of becoming overweight or obese if they eat too much, especially energy-rich food (e.g. fatty and/or sugary foods), and consume too many fizzy drinks, and are not physically active.

Feeding older children and adolescents

All children need three meals and some snacks each day

You can help the older children and youths in your area to be well nourished if you advise their parents to give them three good meals and some snacks each day. Children should have:

- breakfast. This is always important but especially so if the child has to walk a long way to school or work and/or does not eat much at midday. One example of a good breakfast is a starchy food (porridge, bread or cooked cassava) with milk, margarine, peanut butter or cooked beans, and fruit;
- a meal in the middle of the day (see Figure 11). Parents should try to give children a variety of different foods if they take food to school or work (e.g. bread, an egg and some fruit). If children buy food from street vendors or kiosks, they should know which foods give the best value for money (see Topic 2, page 27). If schools in the area provide meals or snacks, you may want to suggest ways to make these as nutrient-rich as possible, for example, by increasing the combination of foods used. If a school has a garden, you may want to make suggestions for increasing the variety of foods grown;
- a meal later in the day. This may be the biggest meal of the day for many children and so it should be a good mixed meal (see Topic 3). Make sure parents realize that fast-growing children are usually hungry children and that they are not being greedy if they want to eat a lot.

Discourage sticky, sugar-rich and salty snacks

There are examples of suitable snacks for older children in Topic 3. Children should know that sweets, sodas and lollies:

- can cause tooth decay;
- can result in an unbalanced diet if eaten in large amounts;
- are poor value for money.

The risk of tooth decay is greatest when foods contain large amounts of sugars and starch that stick to the teeth (sweets/candy, dried fruits, for example) and are eaten often, and when oral hygiene is poor (no or insufficient tooth brushing).

Salty snacks, such as packets of crisps, may also be poor value as they give few nutrients and too much salt, and are costly.

Some children, especially adolescent girls, need to know that it can be dangerous to 'diet'. It is better to stay slim and healthy by eating good foods and being physically active.

Figure 11. School-age children need good food in the middle of the day

Other ways to improve older children's nutrition

- Advise parents to use iodized salt in family meals if it is available. People who lack iodine cannot work or study well. Iodine-deficient girls who become pregnant risk having a baby who is mentally or physically damaged (see Section C, page 9).
- Deworm children regularly, especially those with heavy wormloads. Deworming improves growth and helps to prevent anaemia.
- Make sure that girls and boys know how to avoid unwanted pregnancies (see Topic 5) and HIV/AIDS (which often leads to malnutrition). If a child or youth is HIV+, give advice on feeding (see Topic 10, page 85).
- Teach children about good nutrition in schools and clubs.

SHARING THIS INFORMATION

Before sharing this information with children and their families, you may need to:

1. **Find out.** What meals and snacks are eaten at home by older children and youths. What meals and snacks are provided by schools, employers and vendors. How often children eat. What is eaten for breakfast. How many children do not eat breakfast. What the blocks to feeding older children better are (money, knowledge, time, customs). What the nutrition problems of older children and youths are. What knowledge of nutrition older children have and what they want to learn.
2. **Prioritize.** Decide which information is *most important* to share with parents, other caregivers and children.
3. **Decide whom to reach.** For example: mothers, fathers, other caregivers, older children and youths, and school and youth club staff.
4. **Choose communication methods.** For example: talks, discussions, quizzes, competitions and demonstrations of good meals and snacks, at community and parent/teacher group meetings, and at schools and youth (e.g. Young Farmers) and child-to-child clubs.

Examples of questions to start a discussion

(choose only a few that deal with the information families or children need most)

Why do older children need good meals? What happens if children do not eat well?

Do girls and boys have different nutrient and food needs at different ages?

How often should older children eat?

Why do children need breakfast? Do older children usually eat breakfast? What do they eat? Could we improve breakfasts?

Do children get food at school? Do they take food to school? Are these good mixed meals and snacks? Can we improve the foods children eat during the day?

Do children get a good meal in the evening? Should parents improve these meals? If so, how?

Which snacks or meals do children buy from vendors? Do they know which are good value and which are poor value?

TOPIC 9 FEEDING MEN AND OLD PEOPLE

NUTRITION NOTES

Men and nutrition

Men also need healthy, balanced diets

Like everyone else, men need good meals so they are healthy and active. However, men are usually the better nourished members of the family because:

- they often have more control over the family cash and traditionally may expect and get the biggest and best share of a family meal. For example, they may get a bigger share of meat than women and children;
- they do not have the additional nutritional needs that women have because of menstruation, pregnancy and breastfeeding.

Appendix 2, Table 4, shows that men's energy needs are higher than women's needs, especially if they are doing heavy physical work. But men need less iron than women and girls of reproductive age. So they need less iron-rich food (e.g. meat or liver) than women.

Even so, some men are at risk of undernutrition. The reasons may be that:

- they live alone (e.g. migrant and seasonal workers) and have little cash;
- they do not know how to shop and cook;
- they are single fathers caring for several children;
- the family is very poor or there are severe food shortages;
- the man is ill, or is an alcoholic or on drugs.

Men living alone or who are sole caregivers for children may need advice on how to buy good-value foods (see Topic 2, page 27) and how to make good meals (see Topic 3). They may need recipes that are easy to prepare and advice on food hygiene. Men who are HIV+ need counselling on how to eat well and prevent weight loss (see Topic 10, page 85).

An increasing number of men (and women) need advice on how to prevent obesity or how to lose weight (see Box 19, page 94).

TOPIC 9

Food and care for old people

Eating well helps old people stay healthy and active longer

Old people who eat healthy, balanced diets are likely to stay healthier and active longer. The energy needs of older people are usually less than those of younger people but they need at least the same amounts of protein and micronutrients (see Appendix 2, Table 4).

Old people may have small appetites, so they need nutrient-rich meals

People tend to eat less as they grow older. It is particularly important that old people choose foods that are nutrient-rich so they can get enough nutrients from a smaller amount of food.

Figure 12. Helping old people to eat well

Some old people do not eat enough and so become thin and anaemic because they:

- may have few teeth or sore gums, or are unhappy, lonely or sick;
- are poor or disabled and have no one to help them grow, buy or prepare enough food;
- live in institutions that provide poor meals;
- care for many grandchildren on little money.

Some old people are overweight or obese also because they are unable to be active.

Old people may be able to eat better and be better nourished if you:

- discuss with them how to make easy-to-cook and easy-to-eat meals using a variety of nutrient-rich foods that are good value for money;
- encourage them to take as much exercise as possible. Exercise improves the appetite and helps to keep people healthier and happier, and it helps to prevent overweight and obesity;
- help them get treatment for sore gums and other conditions that reduce the appetite;
- ask people who send money to elderly relatives living alone to arrange for someone to help them to buy and prepare good meals if necessary;
- encourage other people in the community to help needy, lonely old people to cultivate home gardens, shop and cook;
- encourage community income-generating activities that give old people the chance to earn money and feel useful, or that raise money to buy nutritious foods for them;
- advise relatives and people in charge of institutions how to feed old people. For example by:
  - giving small, frequent, good mixed meals to stimulate poor appetites;
  - giving soft foods if teeth are missing or gums are sore;
  - preparing food hygienically to avoid diarrhoea and other infections that may make old people seriously ill.

SHARING THIS INFORMATION

Before sharing this information with families, you may need to:

1. **Find out.** What and where men eat. Whether any groups of men are at risk of undernutrition. If so, why and what advice they need. What old people eat. Whether many old people are undernourished. If so, why. What advice is needed by old people and their relatives.
2. **Prioritize.** Decide which information is *most important* to share with groups or individual families.
3. **Decide whom to reach.** For example: men and old people; people who cook and care for men and old people.
4. **Choose communication methods.** For example: discussions, recipes and cooking demonstrations, at community and farmers' group meetings and at old people's homes.

Examples of questions to start a discussion

(choose only one or two questions that deal with the information families need most)

Why are most men well fed? Are some men undernourished? If so, why?

What advice do undernourished men or men who are sole caregivers for children need? How can we help them?

Why is it important for old people to eat nutrient-rich foods and have healthy, balanced diets?

Are some of our old people undernourished? If so, why?

How can we improve the diets of old people? How can old people help themselves?

How can we help old people who are caring for many children?

TOPIC 10 FEEDING SICK PEOPLE

NUTRITION NOTES

Why sick people need good meals and plenty to drink

Eating well helps to fight infections

Sick people should eat well even if they are not active. They need nutrients to keep alive, fight infections and replace lost nutrients.

Infection often reduces appetite. It also increases the need for some nutrients if:
- nutrients are poorly absorbed by the gut;
- the body uses nutrients faster than usual (e.g. to repair the immune system).

Infections can cause malnutrition. Malnutrition makes infections worse

If sick people do not eat enough, they use their own body fat and muscles for energy and nutrients. They lose weight and become undernourished. Their immune systems may become less effective and they are less able to fight infections.

Sick people often lose or use more water than usual (e.g. during diarrhoea or fever). They need plenty of clean, safe drinks.

Helping sick children and adults to eat well

Feed sick people frequently and give them plenty to drink

Advise families to:

- offer small amounts of food frequently, especially if the person is not hungry. Often a sick person prefers soft foods (e.g. gruel, mashed bananas or soup) or sweet foods. For a few days it does not matter what the person eats, provided he or she eats often;
- give a sick person plenty to drink every 1–2 hours. For example, give boiled water, fresh fruit juice, coconut water, sodas, soup or watery porridge. Or give boiled or soured milk or milky tea unless the person has diarrhoea;
- prepare food and drinks in a clean, safe way (see Topic 4) to prevent food-borne infections

If people are ill for more than a few days, they need a variety of foods to help their immune systems recover and to prevent weight loss (see Box 16). So families should give small, frequent meals that contain a combination of foods (see Topic 3). Adding a little fat-rich food or sugar is an easy way to increase energy without making the meal too big and bulky; including a variety of fruits and vegetables provides micronutrients.

If a young, breastfeeding child is sick, the mother should breastfeed more often. Breastmilk may be the only food and drink the child wants. Advise the mother to express her milk and feed it from a small cup or spoon if a child is too ill to suckle.

In areas where vitamin A deficiency is a problem, children with measles, diarrhoea, respiratory infections or malnutrition often benefit from vitamin A supplements. However, when giving these, health workers should emphasize the need for vitamin A-rich foods as well.

Feeding people with diarrhoea

People with diarrhoea need extra liquids to drink

Children and adults with diarrhoea and/or vomiting lose much water and so must drink frequently to prevent dehydration. Suitable drinks are oral rehydration solution made from packets of oral rehydration salts (from the clinic or pharmacy) or ordinary home-made fluids containing normal amounts of salt, such as soups or rice water.

People with diarrhoea must also eat because food helps the gut to recover and absorb water. Breastfeeding children who have diarrhoea should breastfeed frequently.

BOX 16 • HOW TO HELP SICK PEOPLE EAT MORE

- Offer food every 1–2 hours; give snacks between meals.
- Encourage the person to eat more at each meal.
- Give easy-to-eat foods that the person likes, but include energy-rich and nutrient-rich foods in the meals. For example, give meat, offal, poultry, fish, eggs, and milk foods when possible (adding dry milk powder to porridges and other foods provides extra milk); add extra fat or fatty foods and/or sweet foods, such as sugar or honey.
- Feed when the person has a low temperature, has been washed and has the mouth clean and the nose unblocked.
- Feed the person sitting up (especially if vomiting is likely); feed a child sitting on someone's lap.
- Keep water and food nearby if a person has to stay in bed.
- Never force sick children to eat, as they may choke or vomit.

Feeding people who are recovering

Give extra food during recovery

During recovery from disease most people are hungrier than usual. They can eat more food and quickly regain lost weight. Children can grow faster than normal (catch-up growth). Sick people may have used up their stores of vitamin A, iron and other micronutrients. They need a variety of nutrient-rich foods to fill up these stores again. People can eat more during recovery if they eat extra food at each meal and/or more meals and snacks each day. Breastfeeding children who are recovering from illness should breastfeed more often.

Feeding people living with HIV/AIDS

A healthy, balanced diet helps people who are HIV+ to remain well longer

It is especially important that people living with HIV/AIDS eat healthy, balanced diets. Good diets prevent weight loss and help people to stay healthy longer.

BOX 17 • NUTRIENT NEEDS OF PEOPLE LIVING WITH HIV/AIDS

A WHO expert consultation in 2003 reached the following conclusions regarding nutrient requirements.

Energy needs

- HIV+ adults and children with no symptoms of HIV or other (opportunistic) infections are likely to need 10 percent more energy than non-infected people (see Appendix 2, Table 4, for energy needs of non-infected people) in order to maintain normal weight, activity and growth. HIV+ adults with signs of other infections or AIDS need 20–30 percent more energy to maintain normal weight, and HIV+ children who are losing weight need 50–100 percent more energy.

Protein and fat needs

- There is no evidence at the moment that HIV+ adults or children need extra protein or that fat needs are different from the norm.

Micronutrient needs

- More research is needed on requirements and the role of supplements. It is likely that HIV increases the need for some micronutrients and HIV+ adults and children should have diets that are as healthy and balanced as possible.
  - When pregnant and breastfeeding women cannot have a good diet, they can be given a multiple micronutrient supplement that provides no more than the daily needs of each micronutrient (see Appendix 2, Table 4, for daily needs of some micronutrients).
  - HIV+ children aged six months to 5 years can receive high doses of vitamin A if this is normally given to young children (see Topic 11, page 92).
  - Pregnant women should receive the same iron/folic acid supplements as non-infected women (see Topic 11, page 91).

  Otherwise high doses of micronutrients (particularly vitamin A, zinc and iron) should not be given as these may have negative effects on HIV transmission or progression.

These recommendations may change when more research is reported, so look for the latest information from a reliable source.

Source: WHO. 2003. *Nutrient requirements of people living with HIV/AIDS* (listed in Appendix 3).

People living with HIV/AIDS often become malnourished or more severely malnourished because:

- the HIV infection, other infections and drugs can reduce the appetite, change the taste of food and/or prevent the body from absorbing nutrients;
- they may eat less if they have sore mouths, nausea or vomiting;
- they have increased energy needs because the immune system is working harder than normal (see Box 17);
- they may be tired and depressed, so it is an effort for them to prepare and eat food;
- they may be short of money for food.

Like other sick people, people living with HIV/AIDS who do not eat or absorb enough nutrients use their own body tissues for energy and vital nutrients. They lose weight and become malnourished, and:

- they are less resistant to other infections because the immune system is damaged. This speeds up the downward cycle of additional infections leading to worse malnutrition, leading to additional infections;
- they may absorb smaller amounts of nutrients and drugs (e.g. drugs for tuberculosis, antibiotics, antiretroviral drugs);
- their wounds heal more slowly;
- they feel weak and are less able to work and live a normal life.

It is easier to prevent weight loss during the early stages of HIV infection. Make sure that people living with HIV/AIDS (and their families) know that they should:

- eat healthy, balanced diets (see Topic 3). They do not need a special diet but should have three good meals containing a variety of energy-rich and nutrient-rich foods (including plenty of vegetables and fruits), and they should eat frequent energy- and nutrient-rich snacks each day. People who are already malnourished when they become infected with HIV have especially high energy and nutrient needs. It is essential that they have sufficient food (especially energy-rich foods) and a combination of foods;
- be especially careful about food hygiene. The immune system is under stress so it is important to avoid food-borne infections (see Topic 4);
- take regular exercise because this improves the appetite and builds muscles;
- seek early treatment for infections;
- eat as well as possible when sick and eat extra when they are feeling better in order to regain any weight lost;
- adjust their food intake when they have diarrhoea, a sore mouth, lack of appetite or nutrient malabsorption to make sure they eat enough and choose foods that help recovery.

If people living with HIV/AIDS lose weight, advise them about how to regain it. Discuss how to eat more good foods and encourage physical activity so they rebuild their muscles. See Box 16 on page 83 for ways to help sick people eat more.

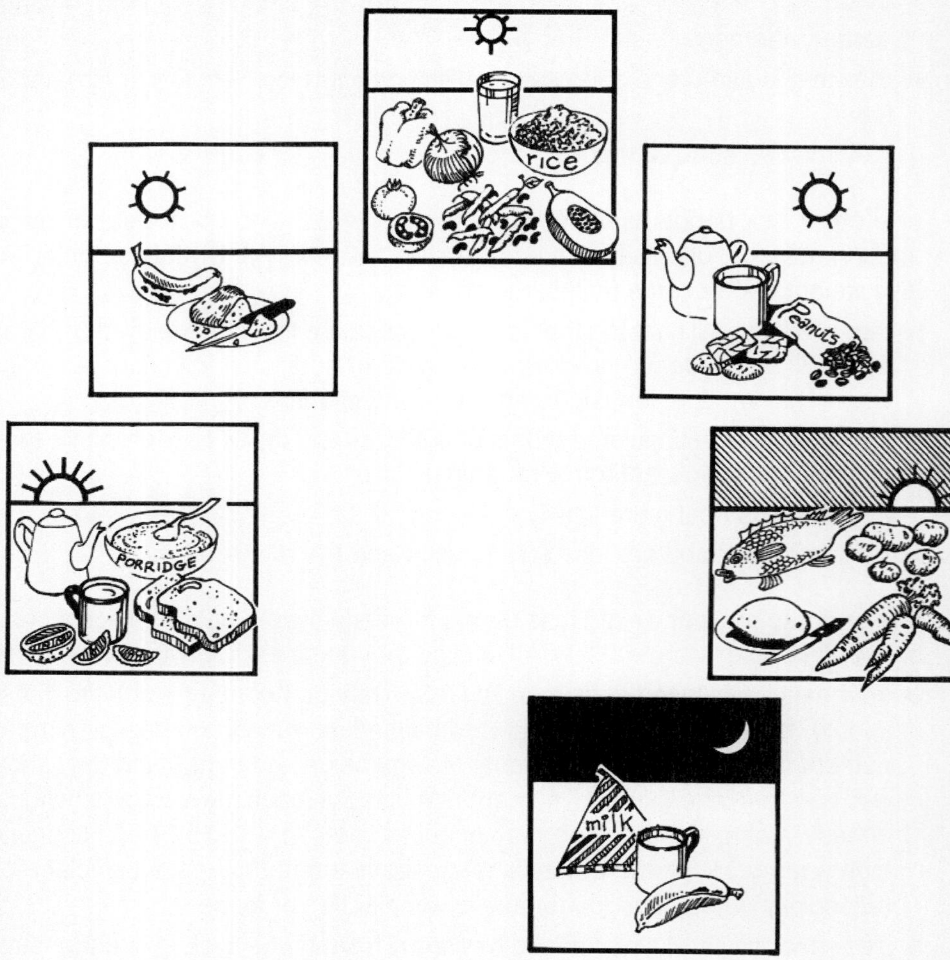

Figure 13. People living with HIV/AIDS need to eat often

There are different interactions between different antiretroviral drugs and different foods. For example, a drug may affect the absorption of a food or a food may affect the absorption of a drug. Health workers may need to consult their supervisors to find out what dietary advice to give people taking antiretroviral drugs (see FANTA/AED. 2003. *Food and nutrition implications of antiretroviral therapy in resource limited settings* listed in Appendix 3).

See FAO/WHO. 2002. *Living well with HIV/AIDS. A manual on nutritional care and support for people living with HIV/AIDS* (listed in Appendix 3) for detailed information on feeding people living with HIV/AIDS, including feeding when there are complications (e.g. diarrhoea).

SHARING THIS INFORMATION

Before sharing this information with families, you may need to:

1. **Find out.** Which foods and drinks are given to sick children and adults (including those with HIV/AIDS). What the local beliefs about feeding sick people and people living with HIV/AIDS are. Who chooses and prepares food for sick people. Who feeds sick people. Whether recovering children and adults are given more food. What the blocks to the better feeding of sick and recovering people are (e.g. caregiver's time).
2. **Prioritize.** Decide which information is *most important* to share with groups or individual families.
3. **Decide whom to reach.** For example: caregivers of sick adults and children; people living with long-term diseases such as HIV/AIDS.
4. **Choose communication methods.** For example: discussions with community and self-help groups and at clinics and during home visits; demonstrations of good meals and snacks for sick people, and people living with HIV/AIDS and their families.

Examples of questions to start a discussion

(choose only a few questions that deal with the information families need most)

Why do sick people need to eat well?

How can we encourage sick people to eat?

Why do recovering children need extra food? How can we give them extra food?

Discuss feeding people living with HIV/AIDS only if a group wants to. If so, do this in a sensitive way. It is usually better to counsel people living with HIV/AIDS and their families individually and, if possible, at home.

Why is it important that people with HIV/AIDS eat well?

Why is it that some people with HIV/AIDS do not want to eat?

Why is it dangerous for people living with HIV/AIDS to lose weight?

TOPIC 11 PREVENTING AND MANAGING MALNUTRITION

NUTRITION NOTES

Undernourished children

Weigh young children regularly and advise on feeding: a healthy child is a growing child

Most children are at greatest risk of malnutrition from the age of about six months (when they are growing fast and breastmilk alone cannot cover nutrient needs) until they are 2–3 years old (when growth slows and they can feed themselves).

Families and health workers can find out if children are well nourished or malnourished by weighing them regularly and plotting their weights on growth charts (see Figure 14). A child may:

- gain weight at the healthy rate, which means the child is almost certainly eating well and is healthy;
- gain weight too slowly or not gain any weight, which signals that something is wrong. The child may be sick and/or not eating enough;
- lose weight, which is a very dangerous sign. The child is not eating enough and is almost certainly ill;
- gain weight faster than the healthy rate, which probably means the child is catching up weight lost during an illness but can also mean that the child has a health problem that could lead to obesity.

A child is *severely malnourished* if there is:

- severe wasting (thinness) *and/or*,
- oedema of both feet.

These children are dangerously ill and need in-patient treatment immediately. Make sure they are kept warm and fed while travelling to hospital.

Undernourished children need frequent nutrient-rich meals

Health workers need to work with the family of a malnourished child to:

- find out why the child is not growing well. Discuss the feeding pattern (amount, variety and frequency of meals), appetite, behaviour and illnesses; examine the child for infections or other medical conditions; try to find the underlying causes (e.g. family food shortages; poor feeding practices; child receives insufficient care). See Introduction, page 9;
- plan together how to help the child. A family will need to:
  - feed the child better. This may mean increasing breastfeeding, improving complementary feeding, feeding more frequently and/or giving more attention during meals (see Topics 6 and 7). Discuss family beliefs on child feeding and blocks to better feeding (e.g. lack of resources, such as food, cash, time or cooking facilities). Then decide together which improved feeding practices the family is able and willing to adopt;
  - take the child for treatment if sick and learn how to prevent childhood infections in the future.

Figure 14. Checking that children are growing well by weighing them often

Health workers should monitor undernourished children's weights closely. If a family is unable to provide a healthy, balanced diet for a child, you may need to give food (enrol the child in a supplementary feeding programme) and micronutrients (e.g. vitamin A and iron) for a while. This must not prevent you from helping the family decide how they can feed the child better. Sometimes a family should be referred to a social worker, agricultural field worker or other community service to help deal with underlying reasons for poor nutrition.

Iron deficiency and anaemia

(also see Introduction, page 9)

Advise anaemic people to eat iron-rich diets and give iron supplements if needed

Signs of anaemia are:

- low haemoglobin (<13 g/dL in men, <12g/dL in non-pregnant women and older children, <11 g/dL in pregnant women and young children and <11.5 g/dL in children aged 5–11 years);
- pale palms and inner eyelids.

The main causes of anaemia are:

- lack of iron. This is often the commonest cause but other nutritional causes include lack of folate, vitamin B_{12} and vitamin A;
- malaria, hookworm infection, other infections (such as HIV/AIDS), heavy bleeding and sickle-cell disease.

People with anaemia:

- need to know how to improve their diets so they get more iron. Improving diets means eating more iron-rich foods (especially meat, offal, poultry and fish) and foods such as fruit that increase iron absorption (see Box 6 in Topic 1). Appendix 1, Tables 1 and 3, list useful sources of iron.
- often need to be prescribed iron supplements and sometimes folate (as folic acid) and other micronutrient supplements – in addition to a good diet. Help people to understand that they must take supplements regularly and for as long as prescribed. Explain the side effects of iron supplements, such as indigestion (which is overcome by taking supplements together with food) and black stools;
- may need treatment for other causes of anaemia, such as hookworm infection, malaria or other parasitic diseases, including schistosomiasis.

Treat all causes of anaemia

Explain to people with anaemia, or their families, how to prevent anaemia in the future by:

- having a diet rich in iron (and vitamin C, if the iron mainly comes from foods of plant origin). Iron supplements may be needed at certain times, such as during pregnancy, but these should never replace a good diet;
- preventing hookworm infection, malaria and other causes of anaemia.

Vitamin A deficiency disorders
(also see Introduction, page 9)

Find out which vitamin A-rich foods are available and promote their use

Lack of vitamin A in the diet weakens the immune system, often causing people (especially children) to become ill and die. If the deficiency is severe, the eye is affected. One of the first eye signs is night blindness (inability to see at dusk and in dim light). There is likely to be a vitamin A deficiency problem in the area if the death rate for children under age 5 years is high (i.e. >50 deaths per 1 000 live births) and/or if many women were night blind during their last live pregnancy (i.e. at least 5 percent).

Families can prevent vitamin A deficiency by:

- eating foods rich in vitamin A (see Appendix 1, Tables 1 and 3). This is the best and only sustainable way to prevent vitamin A deficiency. In order to absorb vitamin A from plant foods well, the meal should contain some fat or oil. If people are unable to obtain a vitamin A-rich diet, it may be necessary to:
  - promote foods fortified with vitamin A (e.g. some oils and fats) if they are available and offer good value for money;
  - give vitamin A supplements to young children and to women within six weeks of giving birth according to national protocols. High doses of vitamin A supplements should never be given to any woman who could be pregnant because they may harm the unborn baby;
  - take children for routine immunizations to prevent infections such as measles. Children with measles often become vitamin A-deficient.

If there are eye signs of vitamin A deficiency, such as night blindness or conjunctival or corneal xerosis (dryness), the person needs urgent medical attention and vitamin A supplements.

Overweight and obesity

Overweight and obese people need less energy-rich foods, a healthy, balanced diet and more exercise

Overweight and obesity (being too fat) are other kinds of malnutrition; in both, the weight is 'too high' in relation to the person's height. Box 18 shows how to determine if an adult's weight is normal.

BOX 18 • BODY MASS INDEX

We use the Body Mass Index (BMI) to decide if an adult has a normal weight or is underweight, overweight or obese.

BMI = weight (kg) divided by height $(m)^2$.

For example, if a woman weighs 50 kg and is 1.5 m tall, her BMI is 50/(1.5 x 1.5) = 22.

Referring to the BMI below, 22 classifies her in the normal weight group.

Weight group	BMI
Underweight	less than 18.5
Normal weight	18.5–24.9
Overweight	25–29.9
Obese	30 and over

People who are overweight or obese are at risk of heart disease, hypertension and stroke, diabetes, certain types of cancers and gallbladder disease. It is most dangerous if a person has a 'fat waist' (the waist is large compared to the hips).

People put on weight when they eat more food energy than they use. This usually is the case when people's normal lives (and work) do not involve much physical activity and their meals contain large amounts of energy-rich foods, such as fats and oils.

Although sugar is not a particularly energy-rich food (see page 21), people who are, or at risk of becoming, overweight or obese should limit the amount they eat. Sugary foods are often rich in fats and they encourage overeating because they are sweet and therefore attractive to many people.

While overweight and obesity is normally seen as a problem of excessive food energy intake only, some health workers do not know that overweight people often also suffer from micronutrient deficiencies (in particular, vitamins A, E and C, and some B-group vitamins) because they often eat poor, unbalanced diets. This is important to note in order to advise overweight and obese people correctly (see Box 19). Not only do they need to reduce their energy intake (and/or increase their physical activity level), but they also must have healthy, balanced diets.

Obesity is a complicated, difficult-to-treat condition in which social norms and values (e.g. fat people are seen as rich people), and psychological factors also play an important role. This makes it more difficult to persuade people to change what they eat and to change their activity level.

BOX 19 • PREVENTING AND MANAGING OVERWEIGHT AND OBESITY

How to help people prevent overweight and obesity

- Explain the risks and causes of overweight and obesity.
- Encourage people to be more physically active whenever possible (at work, play, sports). For example, to walk vigorously for at least 1/2 hour every day or to dig or dance for at least 1/2 hour three times a week.
- Give advice on:
  - what to eat – plenty of fresh fruits and vegetables; lean, instead of fatty meat and fish; wholemeal cereals and pulses;
  - what not to eat – fatty and sugary foods and alcoholic drinks (e.g. 1 litre of beer provides about a tenth of a man's energy needs per day). Fat should supply only about a third of the energy needs; this can only be met if foods rich in fat are eaten only in smaller amounts. Remember that most of the fat in the diet is often 'hidden' in foods like meat, groundnuts, milk and fried foods. The fat in plant foods and fish is usually healthier than fat in meat and milk (see Box 4 in Topic 1).

How to help fat people lose weight

- Explain that:
  - increasing physical activity is essential because regular exercise lowers the risk of heart disease even if there is no weight loss;
  - overweight and obese people must eat less food, especially less high-energy foods and drinks (i.e. fatty and sugary foods/drinks) at each meal. Water, instead of sugary beverages is the better choice. It is also important to eat only when hungry (e.g. avoid eating meals or snacks while watching TV).

People who want to lose weight need a diet containing a variety of foods, especially plenty of vegetables and fresh fruits. Advise them to reduce the amount of beer they drink and to stop eating fatty or sugary snacks; recommend fresh fruits instead.

it is safer to lose weight gradually on a low-energy healthy, balanced diet than on a special slimming diet.

It is difficult for most overweight and obese people to lose weight. They need frequent, sympathetic encouragement. Never laugh at or be rude about obese people – these are serious conditions that need your help.

SHARING THIS INFORMATION

Before sharing this information with families, you may need to:

1. **Find out.** What the common types and causes of malnutrition are, including overweight and obesity. Which types of families are most affected. What the local names and beliefs for poor growth, anaemia, vitamin A deficiency and obesity are. What type of treatment and care is given to people with different types of malnutrition by families and health workers.
2. **Prioritize.** Decide which information is *most important* to share with different groups, families or individuals.
3. **Decide whom to reach.** For example: parents and other caregivers of malnourished children; malnourished adults and their relatives; health staff and volunteers helping at clinics and with community-based growth monitoring activities.
4. **Choose communication methods.** For example: group discussions with community groups and at clinics; feeding demonstrations; individual counselling at clinics and homes.

Examples of questions to start a discussion

(choose only one or two questions that deal with the information families need most)

If many local young children are growing slowly

How can we find out if our children are growing too slowly?

Why do some children grow too slowly?

How can we help these children and their families?

What feeding advice should we share with the families of undernourished young children?

If many children and women have anaemia

Is anaemia (use local name) a problem in this place?

Do you know what causes anaemia? Emphasize the important local causes.

How can we prevent anaemia caused by hookworm, malaria, a poor diet?

Which local foods are rich in iron? How can we improve the amount of iron we absorb from food (see Topic 1, page 19)?

If many people have vitamin A deficiency disorders

What is vitamin A? What happens if a child or adult does not get enough vitamin A?

How can we prevent vitamin A deficiency disorders?

Which local foods are rich in vitamin A?

If many people are overweight or obese

Which health problems are linked to overweight and obesity?

How can we prevent ourselves from becoming overweight?

How can overweight and obese people lose weight? Is it easy?

Appendix 1
NUTRIENTS IN FOODS

Whether or not a food is a good source of a nutrient depends on:

- the amount of nutrient in the food. Foods that contain large amounts of micronutrients compared to their energy content are called 'nutrient-rich' (or sometimes 'nutrient-dense') foods. They are preferred because they help ensure that the diet provides all nutrients needed. This Appendix lists foods that supply useful amounts of different nutrients;
- the amount of the food that is eaten usually.

TABLE 1 • USEFUL SOURCES OF NUTRIENTS

CARBOHYDRATES		
Starches	**Sugars**	**Dietary fibre**
▸ cereals ▸ roots and tubers ▸ starchy fruits ▸ mature legumes	▸ sweet fruits ▸ sugar ▸ honey ▸ sweet foods	▸ wholemeal cereals and roots ▸ legumes ▸ vegetables ▸ fruits
FATS		
Fats high in unsaturated fatty acids	**Fats high in saturated fatty acids**	**Fats high in trans fatty acids**
▸ most vegetable oils (e.g. sunflower, maize, groundnut and olive) ▸ wholegrain cereals ▸ groundnuts, soybeans, sunflower seeds, sesame seeds and other oilseeds ▸ fatty fish ▸ avocados	▸ butter, ghee, lard ▸ whole milk (fresh or soured) ▸ fats from meat and poultry ▸ coconuts ▸ red palm oil	▸ margarine and vegetable ghee ▸ lard/cooking fat

Table 1 • continued

PROTEINS

- breastmilk
- milks from animals
- eggs
- meat and offal of animals, birds and fish
- mature beans, peas and dal
- groundnuts and soybeans
- cereals, if eaten in large amounts

IRON

Easily absorbed

- liver, blood and other offal
- flesh of animals, birds and fish (the redder the flesh, the more iron it contains)
- breastmilk

Poorly absorbed, unless eaten with meat, offal, poultry or fish, or foods rich in vitamin C

- wholegrain cereals, particularly millets and sorghum
- legumes
- amaranthus, spinach and some other dark green leaves

ZINC

- meat and offal
- fish and poultry
- insects

VITAMIN A

- liver and kidneys
- egg yolks
- breastmilk, particularly colostrum
- milk fat, butter and cheese
- whole dried fish (including liver)
- fresh unbleached red palm oil
- orange vegetables, e.g. carrots and pumpkins
- ripe mangoes and pawpaws
- yellow/orange sweet potatoes
- dark/medium green vegetables, e.g. spinach, amaranthus and kale (the darker the leaf, the more vitamin A it contains)
- yellow maize and yellow bananas, if eaten in large amounts

Table 1 • continued

FOLATE

- beans and groundnuts
- fresh vegetables, particularly dark green leaves
- liver and kidneys
- breastmilk
- eggs
- cereals, if eaten in large amounts

VITAMIN C

- fresh fruits, e.g. guava, citrus and baobab
- fresh vegetables, e.g. green leaves, tomatoes and peppers
- breastmilk
- fresh animal milks
- fresh starchy roots and fruits, if eaten in large amounts

TABLE 2 • ENERGY, PROTEIN AND FAT CONTENT OF SOME FOODS

FOOD	% EP	IN 100 g EDIBLE PORTION OF FOOD			
		ENERGY		PROTEIN	FAT
		kcal	MJ	g	g
CEREALS					
Breads, white	100	261	1.09	7.7	2.0
Maize/corn					
► whole, flour	100	353	1.48	9.3	3.8
► refined, flour	100	368	1.54	9.4	1.0
► thick porridge*	100	105	0.44	2.6	0.3
► thin porridge*	100	54	0.23	1.4	–
Millet, bulrush	100	341	1.43	10.4	4.0
Rice, polished					
► raw	100	361	1.51	6.5	1.0
► boiled*	100	123	0.51	2.2	0.3
Sorghum, whole, flour	100	345	1.44	10.7	3.2
STARCHY ROOTS AND FRUITS					
Cassava					
► fresh	74	149	0.62	1.2	0.2
► dried or flour	100	344	1.44	1.6	0.5
► fresh, boiled*	100	149	0.62	1.2	–
Plantains, raw	66	135	0.56	1.2	0.3
Potatoes, Irish, raw	80	79	0.33	2.1	0.1
Sweet potatoes, raw	80	105	0.44	1.7	0.3
Yams, fresh, raw	84	118	0.49	1.5	0.2
LEGUMES					
Beans and peas, dried, raw	100	333	1.39	22.6	0.8
Groundnuts, dried, raw	100	567	2.37	25.8	45.0
Soybeans, dried, raw	100	416	1.74	36.5	20.0
Sunflower seeds, raw	100	605	2.53	22.5	49.0

Table 2 • continued

FOOD	% EP	IN 100 g EDIBLE PORTION OF FOOD			
		ENERGY		PROTEIN	FAT
		kcal	MJ	g	g
ANIMAL FOODS					
Breastmilk	100	70	0.29	1.0	4.4
Cow's milk	100	61	0.26	3.3	3.3
Eggs	88	158	0.66	12.0	11.2
Meat with some fat (goat)	100	161	0.67	19.5	7.9
Chickens/poultry	67	140	0.59	20.0	7.0
Fish flesh, fresh	100	90	0.38	18.4	0.8
Fish flesh, dried, salted, large	100	255	1.07	47.0	7.4
OILS, FATS AND SUGAR					
Edible oils/lard	100	900	3.76	0	100.0
Butter/margarine	100	718	3.00	0	82.0
Sugar	100	400	1.67	0	0

Source: FAO. 1993. *Food and nutrition in the management of group feeding programmes.* Rome.

Notes:

kcal = kilocalorie

MJ = megajoules
(joules are the modern unit for measuring energy. 1 000 kcal = 4.18 MJ)

% EP = Percent edible portion = proportion of the 'as-purchased' weight of food which can be eaten expressed as a percentage

– = trace

* = values calculated. The amount of flour in thick and thin maize 'porridge' varies. These are approximate values only.

TABLE 3 • NUTRIENTS IN SELECTED FOODS

FOOD	RICH SOURCE OF:	USEFUL SOURCE OF:
Cereals	Starch, fibre	Protein B-group vitamins Some minerals
Starchy roots and fruits	Starch, fibre	Some minerals Vitamin C if fresh Vitamin A if yellow
Mature beans and peas	Starch, protein, fibre	B-group vitamins Some minerals
Oilseeds	Fat, protein, fibre	B-group vitamins Some minerals
Meats and fish	Protein, iron, zinc	Other minerals Some vitamins
Liver (all kinds)	Protein Iron Zinc Vitamin A Folate, Other vitamins	–
Milks and milk foods	Fat Protein Some minerals Some vitamins	–
Breastmilk	Fat Protein Most vitamins and minerals except iron	Iron
Eggs	Protein Vitamins	Fat Minerals (not iron)
Fats and oils	Fat	–
Dark/medium green leaves	Vitamin C Folate	Protein Some iron Fibre Vitamin A
Orange vegetables	Vitamin A Vitamin C	Minerals Fibre
Orange fruits	Fruit sugar Vitamin A Vitamin C	Fibre
Citrus fruits	Fruit sugar Vitamin C	–

Source: Adapted from Burgess and others. 1994. *Community nutrition for Eastern Africa.* AMREF, Nairobi.

Appendix 2
ENERGY AND NUTRIENT NEEDS

Use the following table to compare the energy and nutrient needs of different members of the family.

TABLE 4 • DAILY RECOMMENDED INTAKES FOR ENERGY AND NUTRIENTS

SEX/AGE	BODY WEIGHT	ENERGY		PROTEIN	IRON	ZINC	VITAMIN A	VITAMIN C	FOLATE
Years	kg	kcal	MJ	g	mg	mg	mcg RE	mg	mcg DFE
BOTH SEXES									
0–6 months	6.0	524	2.19	11.6	0[a]	1.1	375	25	80
6–11 months	8.9	708	2.97	14.1	9	0.8	400	30	80
1–3	12.1	1 022	4.28	14.0	6	8.4	400	30	160
4–6	18.2	1 352	5.66	22.2	6	10.3	450	30	200
7–9	25.2	1 698	7.10	25.2	9	11.3	500	35	300
GIRLS									
10–17	46.7	2 326	9.73	42.6	14/32[b]	15.5	600	40	400
BOYS									
10–17	49.7	2 824	11.81	47.8	17	19.2	600	40	400
WOMEN	55.0								
18–59		2 408	10.08	41.0	29/11[c]	9.8	500	45	400
Pregnant		+278	+1.17	+6.0	High[d]	15.0	800	55	600
Breastfeeding		+450	+1.90	+17.5	15	16.3	850	70	500
60 and over		2 142	8.96	41.0	11	9.8	600	45	400
MEN	65.0								
18–59		3 091	12.93	49.0	14	14.0	600	45	400
60 and over		2 496	10.44	49.0	14	14.0	600	45	400

Sources: Energy – FAO. 2004. *Human energy requirements.* Report of a Joint FAO/WHO/UNU Expert Consultation. FAO Food and Nutrition Technical Paper Series, No. 1. Rome; Protein – WHO. 1985. *Energy and protein requirements.* Technical Report Series 724. Geneva; Micronutrients – FAO/WHO. 2002. *Human vitamin and mineral requirements.* Report of a Joint FAO/WHO Expert Consultation. Rome.

Notes:

kcal = kilocalorie

MJ = megajoules
(joules are the modern unit for measuring energy. 1 000 kcal = 4.18 MJ)

RE = retinol equivalents

DFE = dietary folate equivalents

These values assume that:

- children are breastfed for at least the first year;
- older children and adults eat small amounts of iron-rich foods (e.g. meat), other animal proteins and vitamin C-rich foods, and large amounts of staple foods such as maize. The bio-availability values used for iron are '10% bio-availability', and those used for zinc are 'low bio-availability';
- adults have moderate physical activity.

a. Full-term babies are born with sufficient iron stores for six months.
b. Amount needed when menstruation starts.
c. Amount needed after menopause.
d. Needs are so high that iron supplements are usually recommended for pregnant women and pregnant adolescent girls.

Appendix 3
ADDITIONAL SOURCES OF INFORMATION

Brown, K. & Wuehler, S. 2000. *Zinc and human health.* From MI.

Burgess, A. & others. 1994. *Community nutrition for Eastern Africa.* From AMREF.

FANTA/AED. 2003. *Food and nutrition implications of antiretroviral therapy in resource limited settings.* From FANTA.

FAO. 2001. *Improving nutrition through home gardening. A training package for preparing field workers in Africa.* From FAO.

FAO/WHO. 2002. *Human vitamin and mineral requirements.* From FAO.

FAO/WHO. 2002. *Living well with HIV/AIDS. A manual on nutritional care and support for people living with HIV/AIDS.* From FAO.

INACG. 2002. *Anemia, iron deficiency and iron deficiency anemia.* From INACG.

IVACG. 2002. *The Annecy accords to assess and control vitamin A deficiency.* From IVACG.

Linkages/AED. 2001. *Essential health sector actions to improve maternal nutrition in Africa.* From Linkages.

Linkage/AED. 2002. *Birth, initiation of breastfeeding and the first 7 days after birth.* Facts for feeding. From Linkages.

Linkages/AED. 2002. *Exclusive breastfeeding: the only water source young infants need.* FAQ Sheet 5. From Linkages.

Linkages/AED. 2002. *Mother-to-mother support for breastfeeding.* FAQ Sheet 2. From Linkages.

Linkages/AED. 2002. *Nutrition Job Aids.* From Linkages.

Linkages/AED. 2002. *Prevention of mother-to-child transmission of HIV in Asia: practical guidance for programs.* From Linkages.

APPENDIX 3

McLaren, D. & Frigg, M. 2001. *SIGHT AND LIFE guidebook on vitamin A in health and disease.* 2nd edition. From Task Force Sight and Life.

WFP. 2000. *Food and nutrition handbook.* From WFP.

WHO. 2000. *Complementary feeding: family foods for breastfed children.* From WHO.

WHO. 2000. *Management of the child with a serious infection or severe malnutrition.* From WHO.

WHO. 2001. *Iron deficiency anaemia assessment, prevention and control: a guide for programme managers.* From WHO.

WHO. 2003. *Nutrient requirements of people living with HIV/AIDS.* Report of a technical consultation 13-15 May 2003. WHO, Geneva.

WHO/FAO. 2003. *Diet, nutrition and the prevention of chronic diseases.* Report of a joint WHO/FAO expert consultation. WHO Technical Report Series 916. From WHO.

WHO/UNICEF/UNFPA/UNAIDS. 2003. *HIV and infant feeding: a guide for health care managers and supervisors* (revised). From WHO.

Addresses for these and other nutrition publications:

AMREF	African Medical and Research Foundation P.O. Box 27691 00506 Nairobi, Kenya E-mail: amrefbooks@amrefhq.org
FANTA/AED	Food and Nutrition Technical Assistance Project Academy for Educational Development 1825 Connecticut Ave., NW Washington, DC 20009, United States of America E-mail: fanta@aed.org
FAO	Food and Agriculture Organization of the United Nations Sales & Marketing Group Viale delle Terme di Caracalla 00100 Rome, Italy E-mail: Publications-sales@fao.org Try for free copies by writing to the Director, Food and Nutrition Division.

INACG	International Nutritional Anemia Consultative Group ILSI Human Nutrition Institute One Thomas Circle, NW, Ninth Floor Washington, DC 20005-5802, United States of America E-mail: hni@ilsi.org
IVACG	International Vitamin A Consultative Group ILSI Human Nutrition Institute One Thomas Circle, NW, Ninth Floor Washington, DC 20005-5802, United States of America E-mail: hni@ilsi.org
Linkages/AED	Linkages Project Academy for Educational Development 1825 Connecticut Ave. Washington, DC 20009, United States of America E-mail: linkages@aed.org
MI	Micronutrient Initiative P.O. Box 56127 250 Albert St. Ottawa, Ontario Canada K1R 7Z1 E-mail: mi@micronutrient.org
SCN	Standing Committee on Nutrition c/o WHO 20, ave. Appia 1211 CH-Geneva 27, Switzerland E-mail: scn@who.int
TALC	Teaching-aids At Low Cost P.O. Box 49 St. Albans Herts AL1 5TX United Kingdom E-mail: info@talcuk.org
	Task Force SIGHT and LIFE P.O. Box 2116 CH-4002 Basel, Switzerland E-mail: sight.life@dsm.com
UNICEF	United Nations Children's Fund Publications Section 3 United Nation Plaza New York, NY 10017, United States of America E-mail: pubdoc@unicef.org

WFP	World Food Programme Via Cesare Giulio Viola 68 Parco dei Medici 00148 Rome, Italy E-mail: wfpinfo@wfp.org
WHO	World Health Organization Distribution and Sales 20, ave. Appia CH-1211 Geneva 27, Switzerland E-mail: bookorders@who.int Try for free copies by writing to the Director, Department of Nutrition for Health and Development

GLOSSARY

Adolescence Period between 10–18 years of age when children are growing into adulthood.

AIDS Acquired immunodeficiency syndrome. A group of diseases caused by HIV.

Anaemia A condition in a person who has a low haemoglobin or haematocrit level. Iron deficiency is the commonest cause. Lack of folate, vitamin B_{12}, vitamin A and other nutrients can be additional nutritional causes. Malaria, hookworm infection, other infections (such as HIV/AIDS), heavy bleeding and sickle-cell disease also cause anaemia.

Body Mass Index (BMI) Measure of thinness or fatness in adults. BMI = weight (kg) divided by height $(m)^2$ (see Topic 11). Normal weight is BMI 18.5–24.9 (see obesity, overweight, below).

Breastmilk substitute Any food used as a partial or total replacement for breastmilk.

Complementary feeding Nourishment of an infant with foods *in addition* to breastmilk or breastmilk substitutes.

Exclusive breastfeeding Nourishment of an infant *only* with breastmilk from the mother or a wet nurse, or with expressed breastmilk, and with no other liquids or solids except drops or syrups consisting of vitamins, mineral supplements or medicines.

Family food security A situation that exists when a family has sufficient safe and nutritious food throughout the year so that all members can meet their dietary needs and food preferences and have active and healthy lives.

Fortified foods Foods with nutrients added to improve their nutritional value. Examples include salt fortified with iodine, and B-group vitamins and iron added to milled cereals.

Healthy, balanced diet A diet that provides an adequate amount and variety of foods to cover a person's energy and nutrient needs.

HIV Human immunodeficiency virus.

Immune system All the mechanisms that defend the body against harmful external agents, particularly viruses, bacteria, fungi and parasites.

Iron deficiency A low level of iron in the blood and other tissues that keeps the body from working properly. It occurs when a person has used up the body's iron stores, and absorbs too little iron from food to cover needs. Iron deficiency is more widespread than anaemia. It is common where the amount of iron in the diet is low, and/or where iron is in a form that is poorly absorbed (i.e. the type of iron found mainly in plant foods).

Macronutrients Nutrients (such as carbohydrates, fats and proteins) required by the body in large amounts.

Malnutrition An abnormal physiological condition caused by deficiencies, excesses or imbalance of energy and nutrients.

Micronutrients Nutrients (such as vitamins and minerals) required by the body in very small amounts.

Nutrient Part of the food that is absorbed and used by the body for energy, growth and repair, and protection from disease.

Nutrition The study of foods, diets and food-related behaviours, and how nutrients are used in the body. People also use the term to describe the food intake of a person (e.g. "He should have better nutrition").

Obesity A condition of being 'too fat'. In adults it means having a Body Mass Index of 30 and above.

Offal Liver, hearts, kidneys, blood, brains and the other non-meat parts of animals, birds or fish that are edible. The redder the offal, the more iron it contains.

Opportunistic infection An infection with a micro-organism that does not ordinarily cause disease, but that becomes pathogenic in a person whose immune system is impaired as by HIV infection.

Overweight A condition of having a weight that is 'too high' in relation to a person's height. In adults it means having a Body Mass Index of 25–29.9.

People living with HIV/AIDS (PLWHA) A general term for all people infected with HIV, whether or not they are showing any symptoms of infection.

Replacement feeding Nourishment of a child who is not receiving breastmilk with a diet that provides all the nutrients the child needs. During the first six months of life this should be a breastmilk substitute.

Vitamin A deficiency disorders (VADD) All the physiological disturbances caused by lack of vitamin A, including clinical signs and symptoms.